Teaching Mathematics Today

2nd Edition

Author
Erin Lehmann, Ed.S.

Forewords
Sharon Rendon, MSCI Math Specialist
Diana Wilson, Ed.D.

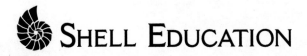

SHELL EDUCATION

Publishing Credits

Corinne Burton, M.A.Ed., *President*;
Kimberly Stockton, M.A.Ed., *Vice President of
Education*; Sara Johnson, M.S.Ed., *Content Director*;
Kristy Stark M.S.Ed., *Editor*; Nicole LeClerc, *Editor*;
Sara Sciuto, *Assistant Editor*; Marissa Dunham,
Editorial Assistant; Grace Le, *Multimedia Designer*

Shell Education

5301 Oceanus Drive
Huntington Beach, CA 92649-1030
http://www.shelleducation.com

ISBN 978-1-4258-1207-2

© 2015 Shell Educational Publishing, Inc.

Teaching Mathematics Today

2nd Edition

Table of Contents

4

Forewords

In *Teaching Mathematics Today, 2nd Edition*, Erin Lehmann outlines specific steps educators can take to transform their classrooms into an environment that is focused on student learning versus a culture of isolation and solely "coverage" of math content. This book is exactly what mathematics educators need to transform classrooms into a dynamic culture with higher levels of learning for everyone involved, where mathematical sense making is valued and expected of each and every student.

Erin draws from research, authors, colleagues, and her own rich experience as a mathematics teacher, coach, and professional developer to present this collection of strategies and instructional practices. She truly believes that mathematics teachers do not serve as the sole source of knowledge and the changing world in which we live requires that both students and teachers learn to learn. Through these reflective practices, both students and teachers become life-long learners and problem solvers that value understanding and are confident to tackle non-routine situations. Mathematics in a technologic world demands a new set of skills and Erin describes these skills and provides support for educators in the teaching of those skills.

I am fortunate to have worked alongside Erin on a regular basis for the past five years and have seen these strategies and practices transform classrooms and support students as they learn rigorous and challenging mathematics. This book will be a resource that will impact your mathematics instruction. You'll hear Erin's coaching voice nudge you to reflect on your beliefs, instruction, and practice and then take risks to implement change. As a reflective practitioner, this resource will provide guidance as you negotiate roadblocks, challenges and plan for future student learning.

I am convinced you will walk away inspired and prepared ready to dive in and make mathematics more meaningful for you and your students.

—Sharon Rendon, MSCI Math Specialist
Coaching Coordinator, CPM Educational Program

The day I met Erin Lehmann I pulled up to North Middle School where the Principal was standing outside greeting tardy students and sending them hastily into the building. I asked him where Erin's office was, she was his full-time math coach, and he said, "Have you met Erin yet? If not, be ready." I was intrigued.

Today I am still intrigued by Erin. She has now expanded her coaching role to the district level and is a proven force in building teachers' instructional capacity in an inquiry-based math classroom. In *Teaching Mathematics Today* Erin offers the next best thing to her being right there beside the teacher. In a concise and comprehensive A-Z, she gives rich explanations and concrete practices that jump from the pages and into action in our classrooms at Rapid City Area Schools.

Teaching mathematics today requires new skill sets which are drastically different than the traditional practices many teachers experienced as students themselves. University mathematics professors struggle to prepare today's math educators for the same reason. Traditional methods teach students to use algorithms to get the answers the teachers want. Today's mathematics instruction must help students gain a deep conceptual understanding of how and why math works and how to apply math in ways that make sense to them.

While many teachers are willing to transform their traditional mathematics instruction to an inquiry-based model, many only reach a superficial level of proficiency. Temptation to continue past practice weighs strong on teachers who find it difficult to let students struggle and persevere in solving problems. They may find it difficult to ask the right questions or may not have the right learning environment to facilitate students' mathematical discourse. If proficiency with inquiry-based instruction is the goal, Erin offers support and constructive examples that will build teachers confidence and competence.

Whether you coach mathematics teachers, whether you supervise mathematics teachers, or whether you are the mathematics teacher, *Teaching Mathematics Today* is a must read.

—Diana Wilson, Ed.D., Director
Office of Curriculum, Instruction and Assessment
Rapid City Area Schools

Introduction

Gone are the days of stand-and-deliver lectures, with the teacher hoping that the students will remember what was said to them—or better yet, that through those lectures students will be able to internalize the complex skills associated with mathematical problem solving. Today's mathematics teachers need to develop trusting relationships with students and create a culture in which students take risks, explore the intricate aspects of problem solving, and participate in mathematical dialogue daily. The classroom is now a place where students can collaborate, explain, question, and justify mathematics.

Teachers should not undertake this shift in practice alone. Being an educator is beyond demanding, and a support system within a school is one way to ease the pressure of being a teacher. One way of easing this pressure is through the use of professional learning communities (PLCs). A professional learning community is a systematic way to help teachers meet the learning needs of all students. In a PLC, teachers work together to impact their classroom practices in ways that will lead to better results for their students, for their team, and for their school (DuFour et al. 2010). It is beneficial for teachers to collaborate with each other when studying state and district standards, creating units, planning summative and formative assessments, analyzing student work and data, and implementing differentiation. Teachers truly are smarter when they focus on the right work together. It takes commitment by teachers and the willingness to learn and study together to build this shared knowledge.

Students need both procedural and conceptual knowledge in order to learn and understand mathematics (NCTM 2000). Knowledge of the procedures and formulas is critical to overall proficiency in mathematics. Also, exploration of the concepts through concrete experiments and manual manipulation is vital to students' overall understanding of the "why" in mathematics instruction. Emphasizing a high degree of procedural proficiency without developing conceptual knowledge is ineffective, and focusing only on conceptual knowledge will not help students achieve in the classroom and in real-world situations. It is necessary to provide focused instruction that moves students from the concrete to the abstract, and then to the application of the concept (Marzano 2003; Sutton and Krueger 2002).

While no single method of instruction has been proven to be the best way to teach mathematics, using research-based designs and procedures has helped educators recognize the best ways to approach learning goals in mathematics (Hiebert and Grouws 2007). Today's mathematics teacher needs to develop a scope and sequence that integrates multiple opportunities for practice prior to an assessment, and time to reteach and revisit those skills that students have not mastered after the assessment. When students are given sufficient practice, they can attempt to use a newly-learned skill in various situations with accuracy so that the skill will be retained (Sousa 2006).

About This Book

Teaching Mathematics Today offers research-based explanations of the strategies that are most critical and highly effective when teaching mathematical concepts, and it promotes the use of best practices. This book reinforces the role of the teacher, who is actively engaged with others, to help students become successful mathematicians. It also offers practical explanations for how to unpack state and national standards to find what students should know, understand, and be able to do.

Teaching Mathematics Today is intended as a guide for mathematics teachers. The book is designed to address grade levels K–12, and it can also be adapted for the various disciplines of mathematics. Here are a few ways educators can use this book:

- Teachers can read the extensive explanations of the mathematics teaching strategies and integrate them into their own lessons.

- Teachers might use this book to improve the effectiveness of mathematics instruction in their classroom.

- A school, mathematics department, or PLC might work through the entire book when streamlining and improving a mathematics program.

- Mentors and coaches can use this book as a guide as they work with teachers, and they can read this book in a study group to discuss best practices in mathematics teaching.

- New and veteran teachers can read and apply the techniques described in the chapters to their current instruction of mathematical concepts.

- Teachers can deepen their knowledge of differentiation to meet the different needs of, and offer access to core curriculum to struggling students and English language learners.

Ultimately, the suggestions in this book will assist teachers in developing their toolbox of research-based teaching strategies to address different learning styles, engage students, and differentiate instruction.

How This Book Is Organized

This book is designed to be read in sequential order. Each chapter focuses on a different aspect of teaching in a mathematics classroom. Here's a brief summary of the topics covered:

- **Chapter 1—Understanding the Role of the Teacher in the Mathematics Classroom:** It is imperative for teachers to develop character trust and competency trust with their students in order to facilitate learning. Students need to wrestle with mathematical concepts in a "productive struggle" and have opportunities to explain and justify their thinking. This chapter provides multiple questioning strategies teachers can use immediately. Because teachers can no longer afford to work in isolation, this chapter also covers collaborative teams of teachers whose members work interdependently to achieve common goals. PLC members are mutually accountable and realize that all of their efforts must be assessed on the basis of results rather than intentions (DuFour et al. 2010).

- **Chapter 2—Managing a Successful Mathematics Classroom:** When strong classroom management techniques are implemented, students have more opportunities for learning. It is important for teachers to have routines and procedures and review them frequently with students. This chapter provides multiple resources for mathematics teachers as they plan for and develop their own classroom management strategies. The planning tools in this chapter will help guide teachers as they implement a workshop model, allowing students to explore and be engaged in mathematics and establishing a sense of community.

- **Chapter 3—Planning for Instruction:** When planning for instruction, it is crucial for mathematics teachers to unpack standards for their grade level and course. Teachers need to attend to the rigor at which the standard is written and hold students to that level of understanding. This chapter offers resources to use when unpacking standards and writing learning targets for instruction.

- **Chapter 4—Implementing Mathematical Practice and Process Standards:** Mathematical practice standards are the "how" in terms of content proficiency. Teachers need to provide opportunities for students to practice engaging in mathematical dialogue in which they defend, explain, justify, and question each other. This chapter provides tools and strategies to reinforce mathematical practices and incorporate them into the mathematics classroom through cooperative learning. When students are actively motivated and busy reaching learning goals, they are also constructing knowledge and moving toward successful mastery of the mathematical content standards.

- **Chapter 5—Building Conceptual Understanding:** Research shows that students who are not successfully mastering mathematical concepts tend to demonstrate slow or inaccurate retrieval of basic mathematical facts, lean toward impulsivity when solving problems, and have difficulty forming mental representations of mathematical concepts or keeping information in working memory (Gersten and Clarke 2007). This chapter offers a step-by-step process to teach students problem-solving strategies along with various vocabulary development activities. The set activity examples are meant to develop independent, competent student problem-solvers.

- **Chapter 6—Assessing Students:** To reach the goal of all students mastering the given curriculum with appropriate instruction, materials, and support, mathematics teachers must use summative and formative assessment strategies "minute by minute and day by day, to adjust their instruction to meet their students' learning needs" (Wiliam 2007). Assessments provide teachers with the necessary data to understand which students are struggling in specific areas of the curriculum and which students need enrichment. This chapter provides strategies, charts, and rubrics for summative and formative assessments, as well as ways to use data to further drive instruction in the classroom.

- **Chapter 7—Supporting Instruction Through Differentiation:** Students should have multiple experiences with topics, allowing them to integrate the topics into their knowledge base (Marzano 2003). However, not all students process the new information in the same way or bring the same skill set to the learning experience. Some students need extra time to process concepts and look at problems in different ways (Sutton and Krueger 2002). Other students may require intense intervention—further teaching or material presented in multiple ways. This chapter provides charts, strategies, and tips for identifying individual student needs and ways to differentiate mathematics instruction to meet those needs.

- **Chapter 8—Integrating Mathematics Across the Curriculum:** Rather than working on subjects in isolation from one another— studying reading apart from writing and apart from math, science, social studies, and other curricular areas—children learn best when they are engaged in inquiries that involve using language to learn and that naturally incorporate content from a variety of subject areas (NCTE 1993). It is important for students to understand that education is not a series of compartmentalized subjects that have nothing to do with one another. Rather, students need to realize that learning is more like a tapestry, where all subjects are woven together to create a broad scope of understanding that is ultimately most useful when all the strands fit together. This chapter provides suggestions and strategies for teachers to integrate mathematics across the curriculum. It includes information about reaching all learners and broadening students' understanding of mathematical concepts.

- **Appendix A:** Teaching mathematics in today's diverse classrooms can be challenging, but it also provides teachers with many exciting opportunities to pass on life skills as well as mathematical knowledge. This section offers general information and suggestions for all teachers and provides additional tips and advice for new teachers.

Conclusion

This book is intended to help teachers navigate today's mathematics classroom. Closed doors are a thing of the past; conversations with fellow educators will help teachers meet the diverse needs of all learners. With a focus on student thinking and learning, this book provides strategies and techniques for teachers to explore and try. Change can only happen through effective teaching by first understanding the role of the teacher, as described next.

Chapter 1

Understanding the Role of the Teacher in the Mathematics Classroom

"Teachers' attitudes and expectations, as well as their knowledge of how to incorporate the cultures, experiences, and needs of their students into their teaching, significantly influence what students learn and the quality of their learning opportunities."
(Banks et al. 2005)

The diversity in today's mathematics classroom requires agility on the part of teachers. Teachers need to not only master mathematical teaching, but they must also demonstrate an innate understanding of the different life experiences students bring to school. According to Megan Tschannen-Moran (2004, 12), "Students who do not trust their teachers or each other will be likely to divert energy into self-protection and away from engagement with the learning task. Moreover, students who do not feel trusted by their teachers and administrators may create barriers to learning as they distance themselves from schools and build an alienated, rebellious youth."

This chapter aims to deepen teachers' understanding of trust and how the different types of trust help teachers navigate the diversity in a classroom. As important as trust is in the classroom, so is an understanding of growth and fixed mind-sets. This chapter outlines how mind-set influences the culture of learning within a classroom.

Student-constructed learning reinforces that learning is an active, meaning-making process, and this process takes time. The strategies listed in this chapter will help teachers provide opportunities for students to do the talking, which leads to understanding. Because planning for success in isolation is challenging, teachers may consider participating in a professional learning community (PLC), a working team of teachers focusing on student learning and increasing the probability of success.

Creating Trusting Relationships with Students

First impressions are critical in developing relationships with students; a student's initial encounter with a teacher can set the tone for his or her engagement and growth the rest of the school year. A teacher should have a warm and welcoming demeanor that invites students to be excited about the adventures they are about to experience in the mathematics classroom.

Students are more willing to take risks and be vulnerable if they trust their teacher and believe that the classroom environment is safe and nurturing. Listening to students is one of the first ways a teacher can begin to build trust. In many classrooms, teachers tend to do all of the talking and they take little time to truly listen to their students. Saying "Tell me more about that" is an effective way to get kids talking. Also, before school, during lunch, in the halls, and at sporting events are all good times when teachers can ask students personal questions and take note of their responses. Students are fairly in tune with body language, so teachers should give full attention to the speaker. True listening is an honest act and not a phony gesture.

Classroom teachers who model both strong character and strong competence are positioned to develop high-trust cultures or classrooms (Moore 2009). Character trust and competency trust, and the ways teachers can develop them, are explained in more detail in the sections that follow.

Character Trust

Character trust includes the behaviors within oneself and needs to be developed and demonstrated each and every day. Students tend to respect the teachers who truly care about them, and they often speak of these teachers highly. Robert and Jana Marzano (2003, 41) state, "If a teacher has a good relationship with students, then students more readily accept the rules and procedures and the disciplinary actions that follow their violations." Without realizing it, educators will also do a lot of teaching outside of their academic content. Teachers are often role models children learn from, so educators have a valuable opportunity to teach positive character traits alongside mathematics. Sometimes this is as simple as greeting students as they walk in the door. Other character traits to model are as follows:

- Truthfulness
- Patience
- Dependability
- Tolerance
- Reliability
- Availability
- Flexibility
- Responsibility

Competency Trust

Competency trust is how a teacher is viewed at doing his or her job. In the eyes of students, this may include teacher preparation, content knowledge, classroom management, and communication with students. If competency trust is high, students are going to trust the teacher, and learning will be the end result. Because the mathematics teacher has created a culture of trust and acceptance, it is okay for students to not know and understand a concept. This high-tolerance classroom, where mathematical conversations allow for questioning and mistakes lead to deeper learning, permits students to be more vulnerable. Teachers strive to get students to talk and collaborate, and they can model the types of positive mathematical conversations they want to hear. An effective way to initiate a mathematical conversation is to ask for a volunteer; students are often eager to do this. Setting the stage for mathematics talk is a way to reinforce this positive behavior again and again. It is important to note when students are having good mathematical conversations. Write down student phrases or peer words of encouragement and share these at the end of class. "How did you get that answer?" will lead to students viewing themselves as mathematicians. This type of positive reinforcement is the structure of a constructive and safe mathematics classroom. If competency trust is low, it is likely there will be management issues, and learning will be secondary to everything else.

Who else should be more excited about mathematics than a math teacher? The enthusiasm for the content needs to be contagious! Each and every day is a new opportunity to express this. Students will be more attentive if the teacher truly loves what he or she is doing and expresses joy while teaching.

Not many people can get excited about the Pythagorean theorem—a math teacher is usually the only exception. Not only is the content exciting, but observing learning is just as exciting. When a student exclaims, "Oh, I get it!" while the teacher walks around the classroom, that teacher should be sure to smile, give a high-five, and celebrate. These are the moments that students will remember down the road.

Mathematics is a series of stories, and when a teacher can relate concepts to students' lives during class, that teacher is making real-life connections that will be memorable for students. The feeling of importance or being noticed is something students look forward to. For example, a word problem could contain the names of students in the class or involve activities students participate in. Perhaps there are fundraisers the class is participating in or mathematics questions students think of that the teacher can use as part of everyday lessons including students.

Finally, the most important thing a teacher can do to create a trusting classroom is to believe in students. "All students can learn math" should be a mantra the teacher repeats daily. Time is the variable in this equation and learning is the constant. Students come to mathematics class at all different levels of readiness. Teachers need to believe *all* students can and will learn mathematics in the classroom each and every day.

Developing a Culture of Learning

The teacher sets the tone for the culture of the classroom. If the teacher is sitting behind a desk as students walk in the room, he or she may not be fostering a welcoming environment. However, if the teacher is up and greeting students enthusiastically, the ambiance is convivial and inviting. The moment students enter the classroom, it is the teacher's responsibility to help them see their greatest potential in a safe and nurturing atmosphere.

In life, students will encounter successes and failures, especially in a mathematics classroom. Intelligence is not fixed, but many students think they are bad at mathematics because it does not come easily to them. Hard work, motivation, and the belief that students can learn can make the difference in learning. A teacher needs to set the stage so that every student can and will learn. Every child has potential, and the teacher should focus on and praise the efforts, risks, and challenges students are undertaking in terms of learning, not necessarily on fixed student intelligence. It is hard to break the stereotype of intelligence being fixed, but when educators promote the fact that learning is fluid, students begin to understand that learning can happen at any given time. In a culture of acceptance and learning, students will have their "ah-ha" moments and educators will celebrate these moments and encourage more risk-taking in learning.

The sections that follow, discuss the importance of student motivation and development of a growth mind-set, praise, responsible actions, expectations, and teacher behaviors as part of the culture of learning.

Student Motivation and Mind-Set

Low student motivation is an issue many educators have observed over the years in the educational system. Fear and intimidation are unfortunately used in many schools today and create a natural resistance in most students. Despite compelling evidence to the contrary, some teachers still believe that fear—fear of failure, fear of an unwanted call home, fear of the teacher, fear of ridicule, or fear of an unpleasant consequence—is a prime motivator for students to do high-quality work (Sullo 2009). In this type of fear-based environment, students will try to be invisible and avoid any and all chance of being noticed, even if this means not asking questions because they do not understand. When fear is instilled in people, adrenalin (the hormone released into the body when we experience fear) inhibits the functioning of the frontal cortex, which is the problem solving part of the brain. When a teacher says, "You really don't know how to add those two fractions. What grade are you in?" the student's thinking stops. When student motivation decreases, it becomes hard for students to think, problem solve, or make good decisions. In the end, using fear and intimidation does nothing to enhance learning.

"How can we motivate students?" is a question educators have been asking for years. Motivation should be an internal drive in everyone, but it seems to be missing in some students. Dangling a carrot or reward in front of students to succeed will work for some, but not for others. As mentioned previously, for years teachers tried using fear as a motivating factor, but that tactic will never achieve internal motivation. According to Bob Sullo (2009), "When students find school and learning to be a need-satisfying experience, they will put working hard and learning into their internal world and will be the academically motivated students we would like them to be." One way to help students experience school as satisfying is to grow their mind-set. If students believe that learning is based more on effort than on intelligence, they will understand the brain is malleable and not fixed.

Carol Dweck's book *Mindset* explains the difference between a fixed and a growth mind-set. Dweck and her colleagues conducted an experiment where they followed several hundred students in New York City during their seventh-grade transition. They measured the students' mind-sets at the beginning of the school year and monitored their grades over the next two years to see how they coped with new challenges. Despite their differing mind-sets, students entered seventh grade with similar mathematical achievement, but their grades split apart in their first term and continued to diverge over the next two years. The students with the growth mind-set (those who believed that intelligence could be developed) significantly outperformed those classmates with a fixed mind-set (Dweck 2008).

A *fixed mind-set* is when looking naturally smart is more important than trying and putting forth effort. A student with a fixed mind-set might see effort as a weakness, and if learning does not happen easily and quickly, he or she gives up. Phrases associated with fixed mind-set might include the following:

- "I can't."
- "This is too hard!"
- "I give up!"
- "This is so stupid."
- "I don't understand decimals, fractions, and percents."

A *growth mind-set* is when a student believes that intelligence is malleable and can increase through hard work and effort. Mistakes happen, and students with a growth mind-set understand that this is when true learning occurs. Students with a growth mind-set look forward to challenges and persevere when confronted with obstacles. Phrases associated with a growth mind-set might include the following:

- "This is hard; I can't wait to figure it out!"

- "Don't stop me now—I want to keep working."

- "I can do this!"

- "Don't give me a hint. I know I can find the solution."

- "I don't understand decimals, fractions, and percents, yet."

To develop students' mind-sets, a teacher could have the students write and reflect on how they feel as learners. The teacher can put phrases like the ones previously mentioned on the board and have students note which ones they tend to say more often. If students lean toward the fixed mind-set, the teacher can remind them that the brain is malleable and it is a choice if they want to have a growth or fixed mind-set. The students can rate themselves daily from a 1 (having a fixed mind-set) to a 5 (having a growth mind-set) and then reflect on why they chose the rating they did. The teacher can ask students to write one goal that they will work on immediately to grow their mind-set. The teacher can also ask each student what they can do to help support these goals.

Another way to grow students' mind-sets is to have them consider Nigel Holmes's graphic on mind-set (see Figure 1.1) and think about which side they resonate with more. The teacher can ask students to give examples of what each mind-set looks like. There are YouTube videos on growth and fixed mind-sets that the teacher can show in class and then ask students to share their thoughts and feelings. Videos for younger students include Tigger and Eeyore from *Winnie the Pooh* thinking about and working through problems. Students can observe Eeyore's fixed mind-set and Tigger's growth mind-set, and then the teacher can ask students what evidence supports these mind-sets. Videos for older students can include famous people who could have gotten stuck in the fixed mind-set due to comments from others but chose to ignore that feedback and continue living their dreams. Also, the teacher can have students share when they have demonstrated a growth or fixed mind-set. It can be enlightening for students to think about the choices they make each day.

Figure 1.1 Fixed vs. Growth Mind-set

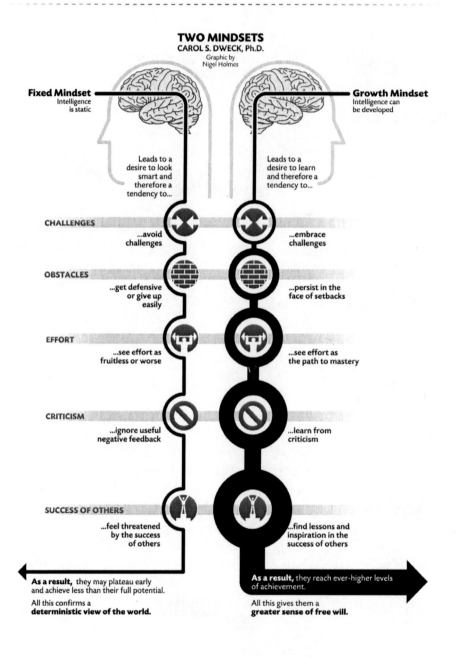

TWO MINDSETS
CAROL S. DWECK, Ph.D.
Graphic by
Nigel Holmes

Fixed Mindset
Intelligence
is static

Growth Mindset
Intelligence can
be developed

Leads to a
desire to look
smart and
therefore a
tendency to...

Leads to a
desire to learn
and therefore a
tendency to...

CHALLENGES
...avoid
challenges

...embrace
challenges

OBSTACLES
...get defensive
or give up
easily

...persist in the
face of setbacks

EFFORT
...see effort as
fruitless or worse

...see effort as
the path to mastery

CRITICISM
...ignore useful
negative feedback

...learn from
criticism

SUCCESS OF OTHERS
...feel threatened
by the success
of others

...find lessons and
inspiration in the
success of others

As a result, they may plateau early
and achieve less than their full potential.

All this confirms a
deterministic view of the world.

As a result, they reach ever-higher levels
of achievement.

All this gives them a
greater sense of free will.

(Reprinted by permission of Nigel Holmes)

Praise

Teachers should be cautious when giving praise to students. Giving praise based solely on intelligence tends to encourage a fixed mind-set. Students may start to take on less challenging activities because the harder activity might be more difficult to figure out and they would in turn appear less intelligent. Instead, praise the processes students choose. For example, if a student is working on finding the area and perimeter of a rectangle and he or she starts by using manipulatives, praise the student on the exploration and perseverance rather than focusing on the right answer. Other areas of praise might include effort, concentration, and the use of different strategies. Praising children's intelligence harms their motivation and their performance (Dweck 2008).

Responsible Actions

Responsible actions can and should be taught in a classroom. Because district and state standards likely include constructing viable arguments and critiquing the reasoning of others, mathematical conversations need to be happening regularly. To maintain a safe culture of learning and conversing, teachers need to be explicit about what responsible actions look like and sound like in a mathematics classroom. One way to address responsible actions is for teachers to brainstorm a personal list of actions that are important to them and to ask students to brainstorm a list they believe are important. For example, a teacher might value promptness, personal accountability, respect, and courage as actions he or she expects to see on a daily basis. Students might value responsibility, effort, and honesty. The teacher can then merge the two sets of beliefs so there is a shared understanding of student and teacher expectations. Once the class has one list, the teacher can create a poster of the behaviors and point out whenever a student exhibits one from the list—for example, "Connie, when you showed Emily how to change the fraction to a decimal, you did it respectfully. That helps us create a community of learners. Thank you." The teacher can give students multiple opportunities to read and reflect on these beliefs, and when the teacher asks students to read and reflect, he or she takes time to do the same.

Expectations

Uncertainty is a scary feeling for many students. As students enter the classroom, they should know immediately what they are going to learn and how it will be measured. To honor a culture of learning, students need to know what is expected of them. Learning targets should be posted so students know what they are going to learn during the day. Each minute is vital in a mathematics classroom. By having the learning target posted, students will not waste time asking, "What are we doing today?" The criteria for success should be listed near the learning target as well. Students will know exactly what they need to do in order to be successful. (Chapter 3 focuses further on learning targets and criteria for success.)

Teaching Behaviors

The actions of the teacher impact the classroom environment. A teacher must cooperatively create an atmosphere of learning for all students. Effective management is a partnership: it depends on collaboration (Routman 2008). Positive actions by the teacher will foster effective teacher-student relationships. In a mathematics classroom, there will be high expectations for deep thinking. Students will be making mathematical mistakes due to the high rigor, which is really a good thing. According to Cathy Seeley (2014, 58), "By creating a classroom environment where mistakes are part of the learning process, we can help students revisit their errors, discuss their thinking, interact with their peers or the teacher, and approach the problem or exercise in a more productive way."

Figure 1.2 shows how teacher actions can be viewed as positive and inviting, so students are excited to talk about mathematics; or negative, so students shut down and are turned off by mathematics.

Figure 1.2 Teacher Actions: Negative and Positive

Negative Actions	Positive Actions
Demands obedience, uses rewards and punishments as motivators	Offers choice and reassurance in a welcoming environment
Uses guilt or shames students into doing work and often criticizes students	Supports independence and expects students to contribute
Does not trust students' ability to learn	Encourages mutual respect and trusts students
Pushes for unattainable goals	Demonstrates patience and encouragement, sets attainable goals

The Teacher as Facilitator vs. Lecturer

Lately, many college programs are emphasizing the role of a teacher as facilitator. Preservice teachers are now taught not to stand in the front of the classroom and talk the entire time, but to let students discover and learn for themselves. Today's teachers encourage collaborative group work. In groups, students tend to listen and share ideas and methods. They need time to wrestle with concepts and discuss challenging problems. New teachers, whose belief is to let students construct meaning, listen and question, pause and paraphrase, will be ready to create this type of learning environment. However, when the going gets tough, new teachers often resort to teaching like they were taught: stand and deliver and put students back into rows. They abandon all student-centered learning and seek silence for a more structured traditional atmosphere. This is likely to happen and it is okay, but teachers should try again the next day. It takes time and patience, but when teachers see students putting their heads together to discuss strategies to solve a logic problem, they realize it is worth it. Students learn to think like mathematicians through collaboration and discovery, and neither is done silently.

Constructivism is a theory of learning based on the principle that learners construct meaning from what they experience; thus, learning is an active, meaning-making process (Glatthorn and Jailall 2009). It takes time and effort to plan for this type of learning, where students make discoveries with real-world problems, but it is essential for all learners. In considering the nature of learning, constructivism involves nine basic principles (Brooks and Brooks 1999; Cornelius-White and Harbaugh 2010; Marzano et al. 1997; Larochelle, Bednarz, and Garrison 1998):

1. **Learning is not a passive, receptive process but is instead an active, meaning-making process.** It involves performing complex tasks that require active use and application of knowledge in solving meaningful problems.

2. **Learning at its best involves conceptual change—modifying one's previous understanding of concepts so that they are more complex and more valid.** Typically the learner begins with a basic or inaccurate concept, and the learning process develops a deeper or truer understanding of the concept.

3. **Learning is subjective and personal.** Learning involves internalizing what is being learned, representing it through learner-generated symbols, metaphors, images, graphics, language, and models.

4. **Learning is situated or contextualized.** Learners carry out tasks and solve meaningful, real-world problems. Rather than doing "exercises" out of context, learners learn to solve contextualized problems.

5. **Learning is social.** Learning at its best develops from interaction with others as perceptions are shared, information is exchanged, and problems are solved collaboratively.

6. **Learning is affective.** Learning is influenced by the following affective elements: self-awareness and beliefs about one's abilities, clarity and strength of learning goals, personal expectations, general state of mind, and motivation to learn.

7. **The nature of the learning task is crucial.** The best learning tasks are characterized by these features: optimal difficulty in relation to the learner's development, relevancy to the leaner's needs, authenticity with respect to the real world, and challenge and novelty for the learner.

8. **Learning is influenced by the learner's development.** Learners move through identifiable stages of physical, intellectual, emotional, and social growth that affect what can be learned and in what depth of understanding. Learners do best when the learning is at their proximal stage of development, challenging enough to require them to stretch but attainable with effort.

9. **Learning at its best involves metacognition.** The learner reflects about learning throughout the entire learning process.

To follow these principles, teachers need to set a balance between lecturer and facilitator. An effective facilitator provides ample opportunities for students to make sense of the mathematics so it can be retained for lifelong learning. Students who are confident in their abilities to do mathematics grow into adults who can navigate complex problems that require analytical thinking. However, students do need direction, and there is a fine line between frustration and disequilibrium. Teachers should step in and offer support through questioning if students seem about to shut down, but not too quickly. Mistakes start new learning, and if students are making errors and are not discouraged, teachers should let them continue with their process. The established atmosphere will invite missteps, and students will be encouraged to offer suggestions and support rather than mockery and ridicule, especially if the culture of a high-trust classroom is established. As Wendy Ward Hoffer (2012, 66) states, "The work of building community is an affirmation of hope for humanity; if we can work together to make meaning of math, what else might be possible?"

Questioning and Feedback

Thoughtful questioning should be the norm in mathematics classrooms. The teacher is already proficient with the mathematical content knowledge, and though it is hard not to be seen as the expert, it is time to put the thinking and learning in the hands of the student. Through thoughtful questioning and effective feedback, teachers allow students to explore the mathematics and make sense of it to ensure retention.

Questioning

In many classrooms, teachers answer hundreds of questions every day. In an inquiry-based mathematics classroom, this number should drop exponentially. Students should ask the teacher a question as a last resort. Most teachers welcome answering student questions and do so automatically. By continually answering questions, however, teachers foster a culture of dependency in which students learn not to think or act by themselves unless they get reassurance from the teacher: "What do I do?" "How do I solve this?" "Is this right?" Once a teacher starts to answer every question, it is difficult to modify that pattern.

The teacher sets the tone immediately during the first few days of school regarding when questions should be asked. The teacher should stress the importance of the collaborative groups and how students need to rely on one another first and foremost. If a student asks a question such as "What do we do when we multiply two negative numbers together?" the teacher can turn to another group member and ask if he or she has the same question. More often than not, the student has not asked his or her group members the question first, so before answering, the teacher needs to make sure the entire group has the same question. Also, the teacher can pause before answering a question and silently count to five. Sometimes students need time to think and respond. Mary Budd Rowe (1986) found that students who were given longer wait times increased the length of their responses by 300 to 700 percent. Teachers might be surprised to find how long students will talk when the teacher does not.

There are times when the teacher should ask the group a question and walk away. This shows that the teacher trusts the group to persevere to find a solution. Teaching students to be dependent on one another, and less dependent on the teacher, helps develop the teacher's role as a true facilitator. As a result, the teacher's role as the adult in the classroom shifts from that of "answer provider" to "facilitator of inquiry," in which the teacher models and reinforces norms or productive discourse and becomes a careful listener (McTighe and Wiggins 2013).

As students are engaged in mathematics, the teacher can have a pre-made list of questions in his or her pocket to use when students need help, and make it a routine to answer a question with a question. Figure 1.3 presents some questions to choose from.

Figure 1.3 Pocket Questions

Questioning Topic	Pocket Questions
Problem solving	What is important?
	What information do you know?
	What strategies could you use?
	What tools could help you?
	Is there another way to think about that?
Getting unstuck/building confidence	What are you being asked to do?
	What do you know about the problem?
	What have you done before that has worked?
	Where did you get confused?
	Is there information you do not need?
	Can you guess and check?
	What do you need to keep working?
Reasoning mathematically	Is this a reasonable answer?
	Would that work all of the time? Tell me why.
	How could you prove that?
	Can you compare answers?
	Can you describe the different solution paths?
Inferencing	What are you thinking?
	What is this about?
	Can you tell me about what your group members think?
	What do you notice? Is it this way all of the time?
Reflecting	How have your thoughts changed?
	How did you get to your solution?
	Does your method seem reasonable? Why or why not?
	Can you think of another strategy that might have worked?
	What was the essential learning for today?
	What are some takeaways from this lesson?

The common practices of "teacher asks question, students raise hands, teacher calls on one student to answer question" and "teacher calls on a student, teacher asks question, student answers correctly or incorrectly" simply do not provide all the benefits of good questioning in a mathematics classroom. In these situations, the teacher hears an answer from only one student while the rest of the students breathe sighs of relief that they were not called on. A classroom observer may note that it is common for teachers to inadvertently call on the same students repeatedly. Students struggling with the content or with the language of instruction might be hesitant to raise their hands or even respond to the teacher's questions. Students' fears of being called on can even keep them from processing the concepts involved in successfully answering the questions. Most important, these teaching practices do not allow all students to demonstrate knowledge and engage in mathematical reasoning and problem solving.

The following questioning strategies provide teachers with a variety of approaches to use, depending on the type of activity/task or instructional setting.

"Everyone Involved" Questioning Strategy

1. The teacher asks a question.

2. The teacher allows students "think time" to process their answers to the question. If necessary, students are allowed to use paper to solve equations and process their own thinking.

3. The teacher directs students to share answers with a partner or in a small group. The teacher encourages students to come to a consensus on the correct answer and to make a case for why their answer is correct.

4. The teacher calls on someone from one of the groups to give the group's answer.

The "Quick Check for Understanding" Cooperative Questioning Strategy

1. The teacher asks a question and displays various responses on the board, one of which is correct.

2. Students work with a partner to solve the problem and determine which answer on the board is correct.

3. The teacher has the partners display their answers.

The "Quick Check for Understanding" Individual Questioning Strategy

1. The teacher asks a question.

2. The teacher allows students "think time" to process their answers to the question. If necessary, students are allowed to use paper to solve equations and process their own thinking.

3. Students record their answers on a small whiteboard or piece of paper. They then hold up the whiteboard or paper so the teacher can see the responses.

Another issue for a teacher to consider when questioning during a mathematics lesson is the level of understanding each student has; the teacher should know the level of mathematical understanding of each student. In addition, if any of the students are English language learners, the teacher should know each student's ability to speak, listen to, read, and write in English and provide specific opportunities for language development to help students make meaning from content (Mora-Flores 2011). When the levels are properly assessed and known, the teacher can use this information for everyone's benefit in a lesson.

The "Leveled" Questioning Strategy

1. The teacher asks a question and calls on a student whose content-readiness level or language acquisition level matches that of the question being asked.

Low Content and Language Complexity Level:

If Meredith earns $5.50 per hour for babysitting, how much money would Meredith earn after 4 hours? How do you know?

High Content and Language Complexity Level:

If Meredith earns $5.50 per hour and her friend Charlie earns $10.00 for the first hour and $3.50 for each additional hour of babysitting, who will earn more money at the end of 4 hours? How do you know? Which family would you want to babysit for and why?

In the first problem, there is only one right answer and the language skills needed to explain the process are very direct. The second problem has multiple solutions and requires a more complex explanation. The number of hours really determines which family would be best to babysit for.

2. Each question asked matches the level or is slightly above the level of the student asked. If a student is just learning English, the teacher should not ask him or her to answer a word problem, but should instead ask the student to find a simple solution using basic recall skills. If a student takes a long time to find a solution for a complex problem, break the problem into easier steps.

The entire class benefits from these differentiation strategies as the students are working with each other to find solutions and participating in the lesson according to their content knowledge or language levels.

Feedback

What is the purpose of a teacher asking students to do a mathematics assignment if it sits on his or her desk ungraded? Teachers need to provide immediate feedback to students and focus on the essential learning goals of the lesson. Feedback can occur individually (e.g., "You are on the right track, but it looks like you are missing a step") or for the entire class (e.g., "Let's take a minute and look at this group's thinking about these angles"). The main goal of feedback is for students to make necessary adjustments in their mathematical thinking. Feedback is not "feedback" unless it can truly feed something (Brookhart 2008).

Using the *Understanding by Design* framework, start thinking of what the end is, plan for instruction, and teach for understanding (Wiggins and McTighe 2005). In each of these steps, the teacher recognizes what the students should know and be able to do. What students are learning should not be a surprise to the students or a guessing game; it needs to be very explicit. With a clear goal in mind, students are more likely to actively seek and listen to feedback (Hattie 2012). When providing feedback to students, the teacher should focus on what they did well or right first before criticizing constructively. The best thing teachers can do for students who fail is to provide them with an honest assessment of why they failed and show them how to do better next time (Jackson 2009).

The teacher's role is to communicate what mastery is before a student starts a project, works on homework, gives a presentation, works in a group, or takes a test or quiz. Rubrics can be created for all of these tasks. Criterion-referenced rubrics describe levels of performance for a particular skill or concept (Dean et al. 2012). Teachers should provide examples at each criterion level to help students better understand the quality of work they are producing. Figure 1.4 shows an example rubric based on mathematical practices and a mathematics problem.

Figure 1.4 Problem-Solving Rubric

Category	Criteria				Comments
	4	**3**	**2**	**1**	
Problem solving	Explanation shows complete understanding of the mathematical concepts used to solve the problem	Explanation shows significant understanding of the mathematical concepts used to solve the problem	Explanation shows some understanding of the mathematical concepts needed to solve the problem	Explanation shows very limited understanding of underlying concepts needed to solve the problem	
Viable argument	Precisely communicates solution to the problem	Adequately communicates solution to the problem	Communicates solution to the problem in a limited way	Inaccurately communicates solution to the problem	
Diagrams/tools	Diagrams/tools are clear and greatly support the concept	Diagrams/tools are clear and easy to understand	Diagrams/tools are somewhat difficult to understand	Diagrams/tools are difficult to understand or are not used	
Overall comments:					

In written feedback, it is a waste of time to provide both a letter grade and comments. Students will look first at the letter grade and often disregard all comments. The grade "trumps" the comments; a student will read a comment that the teacher intended to be descriptive as an explanation of the grade (Brookhart 2008). Choose words that are student-friendly and provide questions to further the student's thinking. Avoid the impulse of solving the mathematics problem for the student—it is for the student to learn and discover through trial and error. Helpful feedback is goal-referenced, tangible and transparent, actionable, user-friendly (specific and personalized), timely, ongoing, and consistent (Wiggins 2012).

Learning how to give effective feedback is difficult, and teachers working in isolation often revert to how they were given feedback when they were students. PLCs can play an important role in effective mathematical instruction. Because a PLC focuses on student learning, feedback becomes integral to the group process.

Professional Learning Communities

One of the keys to an effective mathematics program is having a PLC in place to focus on student learning. A PLC is composed of collaborative teams, focuses on continuous improvement, and is results oriented (DuFour et al. 2010). In a PLC team, educators and administrators should be able to work smarter, not harder. Both small and large decisions have to be made regarding mathematics instruction. When a team is making those decisions, it is more likely that best teaching practices will be strengthened and overarching goals of standards-based instruction will be met.

Depending on the school, district, county, and state, mathematics teams can take many forms and functions. Regardless of the type or function of the team, it is necessary to have administrative support and well-trained, enthusiastic teachers who are committed to increasing student achievement and implementing a well-structured curriculum.

PLC in Mathematics Instruction

A PLC team is most often used at the school level and consists of teachers within a particular grade level or mathematics department, or teacher representatives from all grade levels or mathematics departments. This type of team can make decisions for school-wide mathematics instruction or for a specific grade level's mathematics instruction.

Prior to implementing the mathematics program, the mathematics PLC team should discuss pedagogy and how to come to a consensus. It is recommended that the team meet regularly to review student progress and the timeline for all mathematical concepts. The team should unpack standards, determine essential learning, identify proficiency levels, discuss upcoming lessons, reflect on best teaching practices, and discuss the students' needs in terms of intervention or enrichment. This team can evaluate the flow of content throughout the year or from one course to the next, and the consistency of policies across school mathematics courses.

To ensure successful meetings, the mathematics PLC team may want to make decisions regarding the following issues:

- How decisions will be made within the team
- How to make students' needs the priority for all program-related decisions
- How to develop lesson plan ideas for upcoming concepts
- How to establish curriculum timelines and adjustments based on student understandings
- How course placements can be flexible enough to meet students' changing needs
- When to meet to discuss assessment trends
- When to meet to discuss the effectiveness of the decisions made on the school-wide mathematics program implementation
- How to implement formal and informal assessments in the classroom
- How to interpret data as a way to drive future instruction
- How to handle those students who miss classes
- How to handle those students who are frequently tardy to class
- If and how homework will be assigned
- If and how homework will be graded and reviewed
- What administrative support is necessary to ensure that teachers are effectively implementing the school-wide mathematics program

When planning the curriculum for the school, the team needs to ensure that its curriculum aligns with state and/or district standards. The instructional team uses data to target students' needs and show academic growth. Effective implementation of the curriculum is the driving force behind the overall mathematics program. The sections that follow outline the various PLC configurations.

Cross-Curricular PLC Team

This type of team building takes place mostly at the secondary level. In general, elementary teachers teach across the content areas. They understand that best practices involve integrating the content areas and giving mathematics a function outside of direct mathematics instruction. However, with secondary teachers, this concept is not put into place as often. This type of PLC team could be made up of department representatives across the content areas for each grade level or school-wide. Because learning mathematics is not compartmentalized, different content teachers can work with one another to support the different content areas in authentic learning situations. For example, if a social studies teacher and a mathematics teacher were to pair up, the social studies teacher could research architectural arches and the mathematics teacher could support the social studies content by exploring the geometry of the arch and how the keystone supports the structure.

To ensure successful meetings, teams may want to make decisions regarding the following issues:

- Common curriculum strands between and among content areas
- Projects or activities can be completed for credit in multiple courses using instructional skills necessary in each course
- Effective ways to illustrate connections between content areas to students
- The administrative support necessary to ensure that teachers are effectively implementing cross-curricular instruction

Vertical PLC Team

This type of team discusses various topics of importance to mathematics instruction on a large scale. The purpose is to make vertical connections across the mathematics curriculum and allow schools to consider the broad spectrum of spiraling instruction. For example, since drawing linear equations is an eighth-grade standard, this team would find out how the sixth- and seventh-grade teachers could support this standard or front-load this standard without teaching it to mastery in their respective grade level. This will help eliminate the "inch deep and mile wide" mathematics curriculum that often prevails. A nod to how this connects to ninth-grade mathematics prepares students for that level.

These teams will identify where the essential learning is for all grade levels, so teachers meet the needs of every learner at each specific grade level. Representatives are then able to report back to their home schools to update their colleagues and administrators on the topics discussed during the meeting. For small school districts, this team could consist of mathematics leaders from each school. In larger districts, it might be helpful to only include representatives from schools within each feeder pattern.

To ensure successful meetings, teams may want to discuss the following issues:

- How concepts are taught in each grade level

- How to effectively teach particular concepts

- What curriculum resources are being used at each grade level

- What holes exist in student knowledge, how to bridge the gaps in student knowledge, and how to correct instruction so that future classes of students do not have the same holes

- How to analyze the schools' mathematics data with attention to successes as well as changes or revisions that need to be made for students to be successful

Conclusion

The role of the teacher is ever changing. Establishing a warm and inviting culture nurtures acceptance. Developing a growth mind-set helps students become internally motivated and persevere when learning gets tough. The shift in the teacher's role from lecturer to facilitator fosters an environment where students are at center stage, making sense of the mathematical content with one another. Teachers collaborating within a PLC can ease the decision-making process where student learning is the end goal.

The work of the mathematics teacher can be difficult. The payoff of students viewing themselves as mathematicians and looking forward every day to mathematics class is worth it.

Reflection

1. What are some ways you can develop trust in your classroom? How can you explicitly plan for it?

2. What can you do to help foster a learning environment where the growth mind-set is the norm?

3. What are the differences between facilitating and lecturing?

4. What teams are in place for planning mathematics instruction in your grade level, school, or district, and what function do they serve? If there are no teams in place, which type of team would you like to see implemented, and why?

Chapter 2

Managing a Successful Mathematics Classroom

"Students have more opportunities to learn in a well-managed classroom, and skillful classroom management makes good intellectual work possible."

(Darling-Hammond et al. 2005, 327)

This chapter provides multiple resources for teachers as they plan for and develop their own classroom management strategies.

Math workshop is a way for students to have mathematical conversations where they discover and explore new concepts. Students are making sense of the mathematical tasks as they defend, justify, and explain their thinking. Rather than students sitting in rows, math workshop invites students to work in teams as they support each other's learning journey.

Involving families in student learning provides opportunities for mathematical achievement. According to Anne Henderson et al. (2007, 2), "Many years of research show that involving families contributes to children's academic and social success." Families have a major influence on their children's lives, and it is up to the mathematics teacher to reach out and invite families to help in their child's mathematical learning from day one.

Finally, this chapter discusses classroom management issues and strategies to effectively handle them.

Math Workshop

Samantha Bennett, Lucy Calkins, Nancie Atwell, Laney Sammons, and Wendy Ward Hoffer are authorities on math workshop who have initiated and incorporated this model into the classroom. The workshop model focuses the majority of class time on students, as they think, listen, read, write, and problem-solve to gain a deeper understanding of the mathematical content. The workshop model provides time for students to be actively engaged in the learning process, rather than watching the teacher solve problems at the front of the room. Samantha Bennett (2007) states, "Teaching and learning are so complex...they can be answered only by *multiple* daily opportunities for students to read, write, and talk, and for the teachers to listen to individuals, small groups, and the whole class."

Workshop consists of three components: a mini-lesson, work time, and a reflection also known as Launch, Explore, Summary. These components must comprise the classroom routine each and every day. Students should walk into class knowing the expectations. The structure of the workshop model should not change, only the content. As teachers shift their belief system from direct instruction to inquiry-based learning, rigor and high expectations become the norm. Teachers need to maintain the belief that all students can learn high levels of mathematics.

Workshop provides multiple opportunities for students to explore and engage in mathematics. For example, a teacher might pose the problem, "What are the factors of 24?" Students would collaborate with each other to try to find all of the possible solutions and be prepared to defend their answers. The teacher's role is not to tell students the factors, but to ask "Why?" and "How do you know?" so students take ownership of the learning. This use of time provides for direct instruction, then small-group or independent work, followed by opportunities for metacognition, a chance for individuals to step outside of themselves and notice their own thoughts, growth, and questions (Ward Hoffer 2012). Teachers listen to student discourse; ask fewer, deeper questions; and provide a classroom of supportive learners that fosters student discourse. Ultimately, teachers should trust that students can do the work. They should not step in and rescue students at the first sign of struggle; rather, students should be allowed to wrestle with the mathematics, with teachers continually offering an atmosphere of encouragement and support.

As Wendy Ward Hoffer (2012) explains, "In order for students to devote their time in mathematics class to reasoning and communication as thoughtful mathematicians, we need to offer them something worthy to chew on with their intellectual teeth." Strong mathematics tasks or problems are the keys to a successful workshop. Teachers should take time to look through the textbook to find meaningful tasks. Some rich tasks are provided near the end of mathematics textbooks; other textbooks are set up with strong tasks to begin with. Otherwise, teachers should use their professional learning community (PLC) to create or find mathematics tasks that align with state or district standards. For example, a seventh-grade teacher could give a homework assignment asking students to find unit rates for specific items in a grocery store. The next day, the teacher could have students compare their findings with one another and make inferences.

Once teachers choose a meaningful task in which to engage students, the Launch phase of workshop catches their interest or connects with their prior knowledge. During the Explore phase of the workshop model, the students collaborate in small groups to make sense of the problem and find a solution path. The Summary phase is when students solidify their understanding through whole group presentation. The sections that follow describe each phase in more detail.

Launch

When launching a mathematics lesson, the teacher should make sure it is quick and supports a purpose or overarching content goal to the whole class. This is a good time to invite learners to question mathematics in a real-life situation. For example, if the learning target for sixth graders is "I can find how much to tip at a restaurant," the teacher can show a receipt from a restaurant and ask students to think about what they would need to know to calculate a tip (see Figure 2.1). After students individually brainstorm a list of ideas, the teacher can have them share their ideas with their partner or tablemates. Finally, the teacher can generate a class list of possibilities.

Figure 2.1 Restaurant Receipt

This activity catches students' interest when finding a percent of a quantity as a rate per 100. The following questions might be generated during the class discussion:

- How much do we tip: 10 percent, 15 percent, or 20 percent? Does it matter? What if the service was not good?

- How do we find the percent of a number? Are there different pathways to get to this solution?

- Can I estimate?

During the Launch phase, a teacher can focus on the big concept of the lesson, skills, vocabulary, strategies (new and old), and connections to other problems. The teacher needs to be very conscientious not to reveal too much about a problem, but instead "hook" students into wanting to find a solution or the path to an answer. For the tipping lesson, the teacher could invite a restaurant server to explain to the class what tipping means to him or her and considerations when reporting income tax. The launch is just that: a *plunge* into the essential learning or a rich mathematics task.

Launches should be 7 to 10 minutes at most. This length of time is easy to misjudge, so it helps to set a timer.

Explore

Wendy Ward Hoffer (2012) states, "Work time is the lifeblood of the workshop: This is where the learning happens, when students are released to test the mettle of their minds against a steep challenge, to apply their thinking strategies and use the mathematical practices we have been modeling." The Explore phase is a time for teachers to move from student to student or group to group. By observing, questioning, and providing confirmation, teachers get a good sense of students' level of understanding. Suppose students are exploring relationships between independent and dependent variables—for example, number of hours spent working on homework and test scores. Students discuss which variable is independent and which variable is dependent. Students then explore a data set to find the correlation between the two variables and graph the data. The teacher circulates among groups, encourages students to explain their thinking, and invites other group members to agree or disagree with the correlation, justifying why.

Before students start on a problem, the teacher should decide if they are to work individually, in pairs, in small groups (three members makes a productive group), or as a whole class. The task itself will generally indicate how or if students should work together. The teacher should be explicit as to how students should work, whether it is individually or in groups. For example, if students are working independently, the teacher might emphasize perseverance, ask students what that would look like, and brainstorm ideas of what to do if a student gets stuck. With groups, it is also important to be explicit about expectations. Maybe the day's goal is that all voices are heard and students critique the reasoning of others. Chapter 4 goes into more detail about how to support mathematical practice standards.

In lesson preparation, the teacher identifies a standard, writes a student-friendly learning target, and plans out the activities or questions students will investigate. It is also important to have a list of pocket questions to use during the Explore phase to promote student thinking or to help when students get stuck. Observing students' work will provide insights into misconceptions that might need to be addressed. One way to attend to misconceptions is to use the "catch and release" strategy (Bennett 2007). The teacher calls for all students' attention if it looks as if the whole class is not understanding and says, "Okay class, what I am seeing is many of you are mislabeling the axis. Let us take a minute and clear this up." Taking a few minutes to catch the whole class to talk about misconceptions and redirect students gets everyone back on track and saves time in the long run. The catch and release technique can also be used if students are not on task as productive group members. There is a fine line between direct teaching and clearing up a misconception.

Another way to address misunderstandings is to call for a "math huddle" (Sammons 2010). The teacher asks one person from each group to come to an area of the classroom and talk about the misconception until all learners have a clear understanding of what needs to be corrected. It is then the learners' job to go back to their groups and explain what was discussed in the huddle.

When students are talking about mathematics, it gets noisy. This is an adjustment for teachers if they are used to students silently doing 30 skills problems. The on-task conversation behaviors promote student learning. Teachers can use the 95/5 rule: if 95 percent or more of the conversation is about math, the group is on task. Teachers should always allow for the 5 percent—students are human and social beings. If students are swaying to 90/10 or lower, then it is a good idea to intervene and work with the group on productive mathematical conversations. It is always a privilege for students to work in groups. Sometimes, it takes only one time to lose group privileges for students to appreciate the advantages of input from others.

Students will likely at one time or another complain about having to work in groups or with a specific person in a group. This is the time for the teacher to stress the importance of collaboration and that it isn't always easy, but it is something students will have to do in their mathematics classroom and later in their lives. Figure 2.2 can show students that when they work in groups or teach others, they are learning more than through other methods. Teachers should have high expectations for collaborative group work and be explicit about what that looks and sounds like. Chapter 4 presents a guideline for this structure.

Figure 2.2 Average Retention Rates in the Learning Pyramid

(Adapted from the National Training Laboratories, Bethel, Maine 2003)

Work time is also when a teacher can work one-on-one with a student or with a small group of students. Some students will need more time and attention, and this is when the teacher can meet their needs. For example, in the problem about tipping from the earlier "Launch" section, the teacher might have a group of students who need a hundreds grid or a ratio table. During the Explore phase, the teacher can differentiate by sitting and working with a small group of students, or offer a mini-lesson on parts to whole. The teacher can use alternative or parallel problems before allowing students to start on the original task. Because teachers understand students' abilities, they will know when students will need this scaffold. Chapter 7 discusses differentiation.

Figure 2.3 shows the Summary Planning Tool, which will help prepare for the Summary phase. The Summary phase is when the teacher scaffolds responses so all depths of the task are exposed. An alert and vigilant teacher collecting data on the different levels of learning is the key to an effective summary.

Figure 2.3 Summary Planning Tool

Name of Unit: _____

Investigation: _____

Standard: _____

Learning Target: _____

Student Idea/ Strategy	Performance Descriptors	Rationale for Selection	Sequence for Sharing	Mathematical Ideas to Highlight When Students Share	Teacher Questions/ Actions	Names of Students Who Use the Strategy

(Adapted from Teachers Development Group 2011)

Summary

The Summary phase of the workshop model is critical for solidifying understanding. Unfortunately, this portion of the workshop model is often skipped intentionally or unintentionally, which is detrimental for students. This is the time when students make sense of their learning. Teachers often lose track of time, let students explore right until the bell rings, and miss opportunities for students to solidify their thinking and learning with peers. When students take the time to reflect on solutions as well as methods, compare different solutions or methods, and relate learning to previous situations, they are likely to build a strong understanding of different mathematical concepts and skills. According to Wendy Ward Hoffer (2012), "As you close your workshop, intentionally set aside time for all learners to have the opportunity to reference their good work back to their learning goals." To make sure there is enough time for the summary, a teacher can use a timer—or better yet, the teacher can ask a student to be the timer and allot 5 to 15 minutes for the summary.

The influential article "Orchestrating Discussions" (Smith et al. 2009), identifies five practices that help teachers use students' responses during the Explore phase to advance the mathematical understanding of the class as a whole during the Summary phase:

1. Anticipating student response to challenging mathematical tasks

2. Monitoring students' work on and engagement with tasks

3. Selecting particular students to present their mathematical work

4. Sequencing the student responses that will be displayed in a specific order

5. Connecting different students' responses and connecting the responses to key mathematical ideas

By using the Summary Planning Tool (see Figure 2.3), a teacher will be anticipating the outcome of the summary. When teachers collaborate in their PLCs, they have planned and worked through the problems, finding alternative solutions. In the Summary phase, the teacher facilitates the mathematical discussion, but ultimately the students pose conjectures, make connections, and question one another. This requires a high level of trust in the classroom. It will take the teacher time to get used to listening, pausing, paraphrasing, and trusting that students will engage in high-level mathematical conversations.

Let us take the example problem from the Explore phase about independent and dependent variables. Here is how a teacher could orchestrate discussion and collect data using the five practices:

1. The teacher anticipates that some students will be able to distinguish only the variable and others will be able to make inferences from the data.

2. The teacher notices a group is slightly off task. Standing within earshot, the teacher turns to an on-task group and says, "I notice that all of you are fully participating and not just pretending to be on task." If this subtle approach does not get the off-task group back on track, the teacher can directly address the off-task group.

3. Because the teacher noticed the different levels of student understanding as anticipated, the teacher is ready to select students to present. The students who could only identify the independent and dependent variable will present their work and look like rock stars in front of the class. The teacher has also been on the lookout for high-level thinking and students who made inferences and invites these students to share their findings. This elevates everyone's understanding.

4. The teacher structures the sequence so that first lower thinking is shared, progressing to higher. It is important to be intentional about sequencing.

5. The teacher connects everything back to the learning target.

The teacher is not the knower and holder of mathematical knowledge; rather, the teacher is a facilitator who engages the entire mathematics class to ask thoughtful questions and give meaningful feedback to the student presenters. Once a teacher experiences effective summaries, he or she will easily see the connection piece that is crucial to student understanding.

Presentation of Student Work

The teacher should let students know ahead of time that they will present their work. Wendy Ward Hoffer (2012) calls giving the student a heads up before asking them a question in front of the entire class a "warm calling." This heads up also gives the student an opportunity to say no. Offering students a choice fosters a respectful learning environment. More often than not, students will rise to the occasion (even if they have to bring a buddy for moral support). The teacher will guide the sequence of sharing, but ultimately it is the other students who will ask for more information or probe for clarification.

For these types of conversations to take place, teachers need to model sentence starters. First, the teacher should thank the presenter, and then ask students to state a noticing and wondering statement:

I noticed...

I wonder...

At the beginning, these statements will help engage students, and when students become more comfortable sharing information, they will do so naturally. The web of discussion is the responsibility of the students, not the teacher. If students continually look to the teacher for validation, a level of dependency exists. Notice in the discussion web shown in Figure 2.4, the conversation happens between students. The teacher may ask a question or make a clarification but then removes him- or herself from the conversation and invites another student to join.

Figure 2.4 Discussion Web

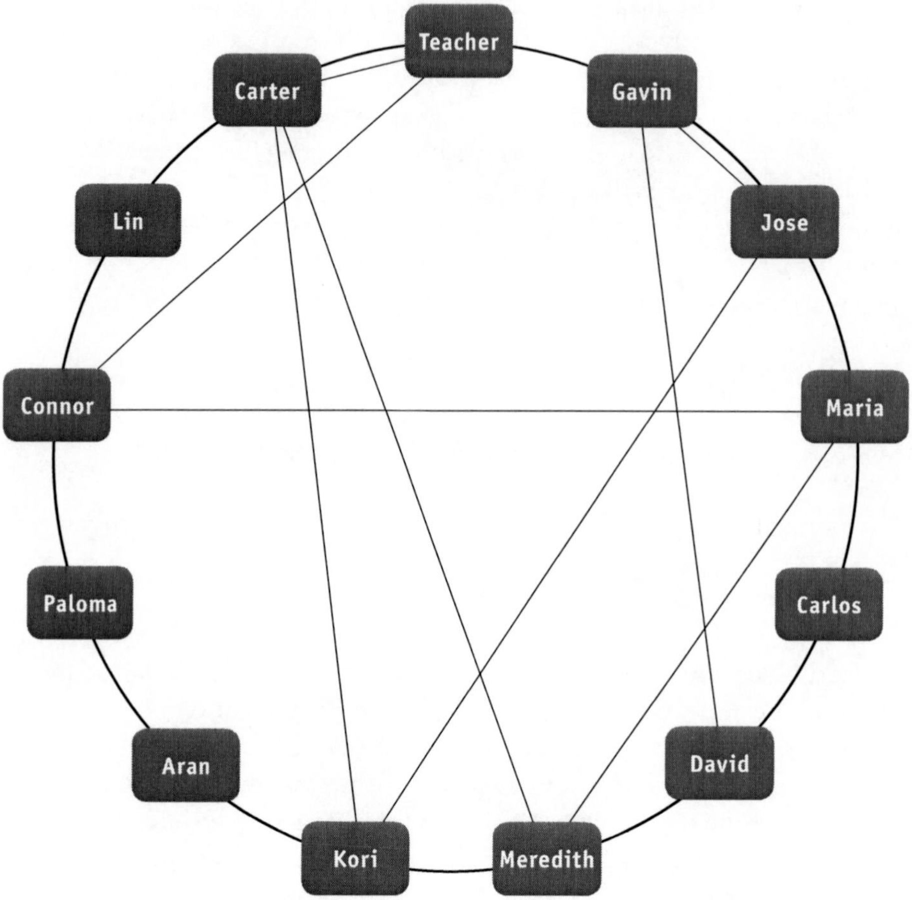

Launch, Explore, Summary Planning Tool

Figure 2.5 shows a planning template that can be used in the workshop model. It is important to take time to fill out this planning tool to help structure each lesson in chunks of time to focus on the essential learning.

Figure 2.5 Launch, Explore, Summary Planning Tool

Unit: _____ Date: _____

Mathematics Standard:	Mathematical Practices:
Learning Target:	

Materials:	Vocabulary:

Inquiry Problem:	Time:
Launch	
Explore	
Summary	

Homework:

Formative or Summative Assessment/Evaluation Plan/Criteria for Success:

(Adapted from Rapid City Area Schools 2010)

Involving Families

It is important to keep positive and open lines of communication with families regarding the education of their children as they progress in understanding mathematical concepts (Seeley 2004). When planning a lesson, teachers should keep their students in mind. However, for a lesson to truly be successful, families need to be informed. Communication and transparency are essential for success in the mathematics classroom. Along with communicating with families, sometimes teachers will need special communication with certain students. It can become a management issue if some students are not on board.

The following suggestions will contribute to success for all students in any mathematics course curriculum.

Parent Involvement

It is the teacher's responsibility to help parents understand specifically how they can assist their children in succeeding in a mathematics course. It is beneficial for all mathematics teachers to involve parents in the process early in the year, in positive ways:

- Establish a good rapport with parents by using effective communication skills.

- Enlist the help of parents as influential allies.

- Use or adapt the Parent/Guardian Letters (see Figures 2.6 and 2.7) as a resource for explaining the significance of students succeeding in the mathematics course and the necessity for students to master mathematical concepts.

- Show parents how they can be vital partners in encouraging their children to remain focused on learning the necessary concepts.

- Make parent communication accessible by translating documents that are sent home if parents speak a language other than English.

Figure 2.6 Parent/Guardian Letter 1

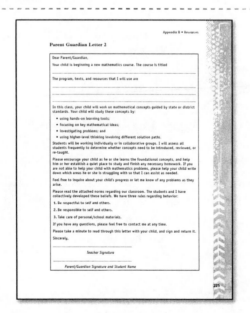

Appendix B • Resources

Parent/Guardian Letters

Parent Guardian Letter 1

Dear Parent/Guardian,
I am excited to have your child in _____ grade!

Some of the big math ideas for this grade are:

In this class, your child will work on mathematical concepts guided by state or district standards. Your child will study these concepts by

- using hands-on learning tools
- focusing on key mathematical ideas
- investigating problems
- using higher-level thinking involving different solution paths

Your child will be working individually, in collaborative groups, or with me. I will assess all students frequently to determine whether concepts need to be introduced, reviewed, or retaught.

Please encourage your child as he or she learns the foundational concepts. Please be available to play math games with your child, provide a quiet place to work on homework, and to continually ask your child to justify how they got their answer because there are multiply ways to get to one solution!

Feel free to inquire about your child's progress or let me know of any problems as they arise.

Please read the attached norms regarding our classroom. The students and I have collectively developed these beliefs. We have three rules regarding behavior:

1. Be respectful to self and others.
2. Be responsible to self and others.
3. Take care of personal/school materials.

If you have any questions, please feel free to contact me at any time.

Please take a minute to read through this letter with your child, and sign and return it.

Sincerely,

_____
Teacher Signature

_____
Parent/Guardian Signature and Student Name

224

Figure 2.7 Parent/Guardian Letter 2

Appendix B • Resources

Parent Guardian Letter 2

Dear Parent/Guardian,
Your child is beginning a new mathematics course. The course is titled

The program, texts, and resources that I will use are

In this class, your child will work on mathematical concepts guided by state or district standards. Your child will study these concepts by:

- using hands-on learning tools;
- focusing on key mathematical ideas;
- investigating problems; and
- using higher-level thinking involving different solution paths.

Students will be working individually or in collaborative groups. I will assess all students frequently to determine whether concepts need to be introduced, reviewed, or re-taught.

Please encourage your child as he or she learns the foundational concepts, and help him or her establish a quiet place to study and finish any necessary homework. If you are not able to help your child with mathematics problems, please help your child write down which areas he or she is struggling with so that I can assist as needed.

Feel free to inquire about your child's progress or let me know of any problems as they arise.

Please read the attached norms regarding our classroom. The students and I have collectively developed these beliefs. We have three rules regarding behavior:

1. Be respectful to self and others.
2. Be responsible to self and others.
3. Take care of personal/school materials.

If you have any questions, please feel free to contact me at any time.

Please take a minute to read through this letter with your child, and sign and return it.

Sincerely,

_____
Teacher Signature

_____
Parent/Guardian Signature and Student Name

225

Once positive contact has been made, parents will be more willing to help a teacher if behavior or academic problems arise. If learning and behavior problems are involved, they are best resolved when teachers and parents work together to examine the context of the problems and deviswa solution. As soon as a problem occurs, it is important for the teacher to call the parents to discuss courses of action to remedy the situation. When problems are discussed, the teacher should always focus the conversation on the behavior and the specific ways in which the parents can assist the teacher in resolving the situation.

Documentation

Documentation is important. Teachers should keep a log of each time parents are contacted to refer to if additional contact or administration involvement is needed. The log also serves as a measure of accountability for the teacher, students, parents, and administration. The log reflects that the teacher notified parents as a preventative course of action prior to giving a failing grade or taking disciplinary action. The log can also show when the teacher contacted parents for positive praise and feedback. For example, a good time for the teacher to make a positive phone call home is when he or she has noticed a student taking the time to explain a mathematics problem to a student who asked if he could just copy it: "I'd love to share with you how I saw your child handle a difficult situation." A student who is struggling with mathematics concepts would be another reason to call home. This documentation could be used in the future if the student is identified for specific mathematics interventions.

The Teacher/Parent Contact Log shown in Figure 2.8 can be used for this purpose. The teacher can quickly record the date of the contact in the first column and the time of the contact in the second column, indicate the topics that were discussed (absences/tardies, grades, conduct, participation), and record to whom the teacher spoke. The teacher should add any final notes regarding action steps or follow-up in the last column. By keeping one log for each student, the teacher is able to quickly reference the reasons, frequency, and dates of contact.

Figure 2.8 Teacher/Parent Contact Log

Student Name: _____ Class/Period/Section: _____

Phone Number: _____ Parent/Guardian Name:_____

Date	Time	Call or Email	Topics: (Check all that apply)					Spoke To	Comments
			Absences/Tardies	Grades	Conduct	Participation	Other		

Student Involvement

A culture of learning depends on a safe classroom where mathematical conversations flow. To ensure safety, it may be necessary to create a contract with some students. This is a democratic process that allows students to have ownership of their roles and their behavior in the classroom. Contracts can be developed after students and teachers agree on the rules and procedures of a classroom. Furthermore, discussions can be initiated regarding both teacher and student expectations. Contracts can be used in the classroom setting or as a school-wide policy. The Social Contract in Figure 2.9 can be used for this purpose, or a school may develop its own contract.

Figure 2.9 Social Contract

Student Name: _____

Teacher Name: _____

Class/Period/Section: _____

1. What ways can a teacher help you?

   _____

   _____

2. What does respect look/sound like?

   _____

   _____

3. What do you expect from your teacher?

   _____

   _____

4. How will you hold yourself accountable as a learner?

   _____

   _____

5. If problems arise, how do you want to resolve them?

   _____

   _____

6. Explain what you can do to help make sure the classroom is a good learning environment.

   _____

   _____

The Student Letters shown in Figures 2.10 and 2.11 can also be used to explain to students the significance of participating and succeeding in the mathematics course and the purposes for mastering the concepts.

Figure 2.10 Student Letter 1

Student Letters

Student Letter 1

Dear Student,

Welcome to _____ grade! I am really excited you are here.

The big math ideas for _____ grade are:

I am here to help you become a great mathematician. It is important that you come to school each day ready as a learner. That means you should

- be curious, open-minded, and thoughtful
- be accepting and patient of others (not everyone learns the same way you do)
- be respectful of the risks people take (listen carefully to others and do not interrupt)
- be willing to take risks (share ideas and other ways of finding solutions)
- be responsible for your actions
- help others be responsible for their actions

As a mathematician, it is very important that you participate in the lessons. This will help you grow as a learner.

Please let me know if you are struggling and need more practice or time. We will work together. We will have fun and learn math at the same time!

The rules our class made are:

1.

2.

3.

4.

5.

Sincerely,

_____
Teacher Signature

_____
Parent/Guardian Signature and Student Name

226

Figure 2.11 Student Letter 2

Student Letter 2

Dear Student,

You are starting a new mathematics course. The course is titled

The books and resources that you will use are

This course will help you develop as a mathematician. It is vital that you take an active, meaning-making stance as a learner. You will be required to do the following:

- Take a learner's stance (be curious, open-minded, and thoughtful)
- Be accepting and patient of others (not everyone learns the same way you do)
- Be respectful of the risks people take (listen carefully to others, do not interrupt, and be aware of body language)
- Be willing to take risks (share ideas, add to brainstorming, and suggest alternative strategies)
- Be responsible for your actions
- Help others be responsible for their actions

As a mathematician, it is essential that you participate in the lessons in order to grow and develop as a learner.

You are responsible for letting me know if you are struggling with some of the concepts and need more practice or time. We will work together to understand, to make sure you are successful.

The norms our class developed are as follows:

1.

2.

3.

4.

5.

If you are not able to follow these norms, what are some alternative solutions we can agree upon? _____

Sincerely,

_____
Teacher signature

_____
Student signature

227

Common Classroom Management Issues

Classroom management refers to all the structures, routines, and procedures that teachers specifically plan for. When a classroom is running smoothly, students are able to collaborate with one another and focus on the mathematics with little to no wasted time. However, when teachers do not plan effectively for learning, the result can be a chaotic environment in which mathematical thinking and learning is challenging. The following sections provide structures for teachers to think about as they plan their lessons.

The Physical Environment

The physical classroom environment is often the outcome of a teacher's outlook and influences from educational pedagogy. The classroom setup also has an effect on the way students, parents, and others feel as they enter the room. Teachers need to examine the goals they have for student learning and then set out to arrange the classroom in a way that will facilitate the types of learning activities that will promote student success.

As teachers walk through the classroom they have set up, they can consider these questions:

- Can students easily access their desks, materials, textbooks, and manipulatives? How will they get paper, pencils, or other supplies before or during mathematics lessons?

- Is the classroom set up for teacher lecture only, or is there some allowance for cooperative learning and interactive activities?

- Is it easy to see the board or overhead projector from every desk?

- Are the walls cluttered? Can students find information they need to help them review major mathematical concepts?

- Are the walls bare? Is there any support as students try to remember key vocabulary, formulas, and ideas specific to mathematics?

- What colors have been used in the classroom? What general emotion do they convey?

- Is there a place to clearly display classroom rules, norms, and daily lesson objectives?

- Is there a place where student work and efforts are celebrated?

- Is the general look of the classroom clean and organized or messy and cluttered?

- Is the physical space easily adaptable for direct instruction, cooperative group activities, centers, and quiet, independent practice of skills?

The answers to these questions will help teachers determine what creates a warm and welcoming environment. Students are more willing to collaborate and share with others when their surroundings are inviting. According to Lonnie Moore, "Trust and communication work synergistically together. Better communication results in higher trust, and higher trust results in easier communication" (2009, 104). Teachers should arrange the classroom so trust flourishes and communication flows.

Interactive Learning Activities

Many teachers hesitate to use interactive learning activities because they fear the chaos that may ensue. Others enjoy allowing groups to work together so they can sit at their desks and correct papers without monitoring the work that is being done. These are two extremes to approaching interactive learning in a mathematics classroom; neither of them is recommended. Learning is not passive, but a true meaning-making process. Teaching cannot be passive either. A teacher must be actively present when students experiment and solve for "why" in a classroom. A constructivist approach to learning requires an engaged, energized teacher.

The following guidelines will help teachers successfully incorporate interactive learning activities into a mathematics lesson:

1. Plan and prepare for interactive activities before students enter the classroom. All materials need to be assembled and ready to use.

2. Model and role-play the procedures expected of students. As students become more familiar with the procedures of the activity, more in-depth mathematics content can be introduced.

3. Remind students of the following every time they participate in an interactive activity:

- The activity's purpose
- The expectations for student conduct
- The consequences or alternative activity for not meeting the expectations
- The procedures for the interactive activity

4. Follow through. Be consistent and make sure to honor interactive learning. Limit direct teaching because this interactive time is when students will make sense of the mathematics.

5. Be the most observant person in the room during an interactive activity. Constantly walk around the room in close proximity to all students, monitoring the responses and the student work, posing questions, pushing for deeper thinking, and providing feedback.

6. If the first attempt is not successful, try again! It takes time for students to become familiar with the procedures and the atmosphere.

The following interactive activity involves pairs of students. The teacher should review expectations of group work and be explicit about what he or she should see and hear as students work together.

To review ordered pairs, students can play ordered pair battleship.

1. Each student places a 2, 3, 4, 5, and 6-point ship on his or her own coordinate plane (shown on the next page). Ships may be placed horizontally, vertically, or diagonally.

2. Students work with partners and guess where their partners placed ships on the coordinate plane without looking at their partners' paper.

3. The goal is for students to sink their partners' ships. Students guess an ordered pair and their partner will say "hit" or "miss."

4. Students need to keep track of hits with red *X*s and misses with red *O*s when their partner guesses, and green *X*s and *O*s for their own guesses of hits and misses. If students get a "hit," they may go again. If not, it is their partners' turn.

5. The winner is the first person who sinks all five ships.

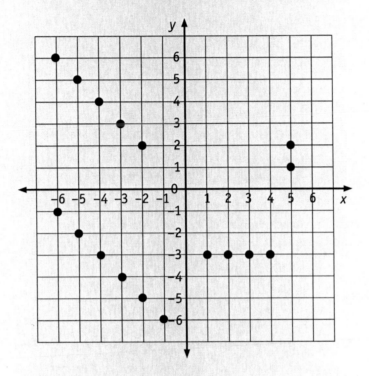

Mathematics Classroom Procedures

Every successful classroom needs established norms, procedures, celebrations, and consequences. Every teacher has a different way to effectively set these up in the classroom. Classroom norms are a shared understanding of what is expected from the class. These norms should be established early in the school year and clearly posted for everyone to see, in addition to celebrations and consequences of following, or not following, them.

What are often not as clear as the norms, however, are the procedures. The block of time designated for mathematics instruction each day (or each mathematics period) has its own procedures, which are the typical sequence of events in a mathematics classroom. Mathematics teachers need to make decisions on the procedures specific to the mathematics instructional time and clearly explain the expectations to students, for example:

- What will students do when they are finished with an assignment during independent practice?

- Are students allowed out of their seats for any reason during independent practice or during cooperative group activities?

- What noise level is acceptable for the classroom or a particular type of activity?

- How will students get paper or sharpened pencils if they need them?

- How will the teacher distribute manipulatives or calculators? How will these things be collected after use?

- Is restroom use or drinking water from a water fountain permitted during instructional time?

- What should a student do if he or she needs assistance solving a problem?

- During group work, may a group discuss ideas and questions with other groups?

- How will homework be assigned, collected, and graded?

- How will students access correct answers to homework or independent-practice problems?

Procedures, like norms, need to be reviewed often. Going over procedures once will not ensure students know what is expected of them. Teachers should take the time to practice different procedures so students get used to the routines in their mathematics classroom. The more teachers review and practice routines, the more students will understand how to manage themselves within the classroom. Having procedures posted on classroom walls is another friendly reminder of what is expected of students. For students to take more ownership in these procedures, teachers can have them create the posters that will be displayed in the classroom.

Conclusion

Planning is key for a high-functioning mathematics classroom. Structures help students know what is expected of them as learners, and routines help minimize classroom disruptions. Families play a critical role in a student's mathematical success. Teachers, students, and families need to work together to reinforce the learning that is happening inside and outside of school. As teachers consciously plan for instruction, students will always be at the forefront of the mathematical thinking and learning.

Reflection

1. Fill in the *Launch, Explore, Summary Planning Tool* (see Figure 2.5) when planning a lesson. Which parts were easy to fill out? Which parts took more time?

2. In what ways will you involve parents and guardians in your classroom? How will you communicate with them? How often? What are the benefits of open channels of communication?

3. Take time to write down what you want/expect from students. How can you help that happen? What is your role in creating a positive learning environment?

Chapter 3

Planning for Instruction

State and district standards provide a coherent focus for each grade level. Before the introduction of the Common Core State Standards (CCSS), states across the country had many different adaptations of state standards that varied in number and how they addressed depth of knowledge. Standards should emphasize conceptual understanding, which allows students to construct their own learning that will eventually prepare them for college and careers. In his article "Building on the Common Core," David Conley (2011, 18) states:

"The ideal result of standards implementation will be to move classroom teaching away from a focus on worksheets, drill-and-memorize activities, and elaborate test-coaching programs, and toward an engaging, challenging curriculum that supports content acquisition through a range of instructional modes and techniques, including many that develop student cognitive strategies."

Any set of standards should tell the "what" of instruction, but it is the teacher's responsibility to structure the "how." Ultimately, standards provide a clear continuum for all K–12 learners of mathematics.

All standards-based instruction helps students to be at the forefront of worldwide technological, scientific, and mathematical advances, trade, and development. Standards provide mathematical accountability so students develop the necessary skills to continue their learning processes in later grade levels, and higher mathematics in college and beyond (Dean and Florian 2001). Furthermore, standards build on each other. For example, the CCSS for Mathematics (CCSSM) were written from a

Progressions Document funded by the Brookhill Foundation, which was based on children's cognitive development and the coherent structure of mathematics. A fifth-grade teacher can plan with a reasonable expectation of where students ended in their fourth-grade instruction based on the progression of these standards. This will help the fifth-grade teacher plan a precise beginning with an awareness of what standards need to be covered throughout the year. While there will always be students who struggle with the various concepts, standards-based instruction ensures that students have at least been introduced to each important mathematical concept for their grade level and assessed on the levels of mastery of that concept.

This chapter explores the importance of knowing and understanding how standards are written and how to unpack them. Many resources are provided for educators to use when unpacking standards or writing learning targets. All standards are written at different depths of knowledge, and it is important to teach and assess at these same levels. This information will be useful when planning for short- and long-term instruction.

Unpacking Standards

Teachers need to collaboratively work together to figure out the "how" behind the standards. Teachers should read through their state or national standards for their grade level and discuss the technical meaning with their professional learning community (PLC). They will want to take time to unpack the standards using a standard unpacking template like the one shown in Figure 3.1 to identify specific learning outcomes for each domain and cluster.

Figure 3.1 Standard Unpacking Template

Domain/Cluster:			
Standard:	Concepts students should know (list **nouns**):	Skills students should demonstrate (list **verbs**):	**Level of thinking:** **Webb's Depth of Knowledge (DOK)** 1 – Recall 2 – Skill/concepts 3 – Strategic thinking 4 – Extended thinking
Student-friendly learning target(s):			
Key vocabulary:			

The last column in the Standard Unpacking Template, "Level of thinking," is vital in helping maintain the authentic structure of the standard. Too often, teachers take a standard and, after unpacking it, keep it at the recall level to easily measure mastery. Norman Webb's Depth of Knowledge (DOK; Webb 2005) is just one example of a framework tool to examine levels of thinking. Bloom's Taxonomy for Learning and Marzano's Taxonomy are frameworks that are commonly used as well. Teachers can use any of the following frameworks when looking at levels of understanding.

Depth of Knowledge Levels:

- DOK 1: Recall
- DOK 2: Skill/concept
- DOK 3: Strategic thinking
- DOK 4: Extended thinking

Bloom's Taxonomy for Learning (revised; Anderson and Krathwohl 2001):

- Remembering
- Understanding
- Applying
- Analyzing
- Evaluating
- Creating

Marzano's Taxonomy (Marzano 2000):

- Level 1: Retrieval
- Level 2: Comprehension
- Level 3: Analysis
- Level 4: Knowledge utilization
- Level 5: Metacognition
- Level 6: Self-system thinking

A level 1 question or recall question could be as follows:

Solve 86 × 94

A student would use a strategy to multiply and find a solution.

A level 3 or analyzing problem could be as follows:

Connie solved the following multiplication problem this way:

86 × 94

80 × 90 = 7,200

6 × 4 = 24

7,200 + 24 = 7,224

Look at how Connie solved the multiplication problem. Do you agree or disagree with Connie's strategy? Explain your reasoning.

Writing Learning Targets and Criteria for Success

After unpacking a standard, the next step is to write student-friendly learning targets. "I can" and "We can" statements foster positive language that communicates what students will learn. If students are unsure of the language—for example, if they encounter "I understand $p + q$ as the number located a distance q from p, in the positive or negative direction depending on whether q is positive or negative"—they are likely to shut down or choose not to learn at all. This type of language is more suitable for an instructional objective than a learning target. A student-friendly learning target could be, "I can find the absolute value of a number." The language is short, concise, and to the point.

Because learning targets are intended to be short and concise, with a quick focus on what students will learn by the end of the lesson, they give a daily purpose for students, telling what students are going to learn and how deeply they will learn it. The most effective teaching and the most meaningful student learning happens when teachers design the right learning target for the day's lesson and use it along with their students to aim for and assess understanding (Moss and Brookhart 2012).

Learning targets can be written in four different ways (Stiggins et al. 2007):

- **Knowledge:** Know and understand a concept or skill
- **Reasoning:** Use and apply knowledge
- **Skills:** Based on student performance
- **Product:** Based on tangible products

Teachers sometimes use activities as learning targets, but activities are the means to achieve the learning target. The answer to the question, "What do I want students to know and be able to do?" should suffice as the learning target. "I can use fraction strips to add fractions" is an example of a poorly written learning target. The activity or task in the classroom is using the fraction strips, whereas the actual learning is adding fractions. An improved learning target is, "I can add fractions." Teachers should take the time to work with their PLC members to develop learning targets. Students will also need to know what the criteria for success are to achieve the learning target. What will it look like if students have hit the learning target, and how will the teacher know? First, the teacher should be aware of what he or she is measuring. The teacher can use rubrics, exemplars, self-reflection tools, or models to help clarify the criteria for success.

Learning targets should be seen and heard. Teachers should write the learning target so it is visible the entire class period. Also, they can have students read the learning target and paraphrase what they think it might mean, and discuss answers in small groups or as a class. Teachers can have students record the learning target as a focus of what they are going to learn and have students reflect at the end of class. Meaningful sharing requires that teachers use the learning target with their students and students use it with one another (Moss and Brookhart 2012). Students can take a self-assessment on whether they attained the learning target.

Students will usually be honest in their answers, and their responses will give the teacher crucial information about where he or she needs to go instructionally for the next lesson. Exit slips are an efficient way to obtain this information.

Aligning Instruction with the Standards

In order for students to have the opportunity to understand concepts that are grade-level and developmentally appropriate, mathematics lessons should be based on standards. Because it is necessary to teach to the content standards and not "to the textbook," teachers should use high-quality problems or tasks that ask students to problem-solve and explore, rather than relying on rote memorization of skills or procedures. If it is not the case that curriculum and standards are aligned in a teacher's school or district, the teacher will need to find tasks or problems that have an emphasis on problem solving for conceptual understanding with embedded skills or procedures.

Here are some guidelines for rich tasks that will help deepen student understanding:

- Focus on conceptual understanding.
- Provide opportunities to use and refine skills when solving problems.
- Include engaging questions.
- Offer various paths to the solution.
- Require higher-level thinking.
- Connect the mathematics content to disciplines outside of mathematics.
- Provide a few rich problems as opposed to many problems.

The CCSSM, for example, require sixth-grade students to display and summarize data sets. Instead of giving students prescribed data and asking them to make inferences, teachers could structure a lesson in which students collect data, display it, and draw conclusions from their findings. For example, students could collect data from classmates asking how many siblings a student has, who was born in what month, or how much sleep a student gets at night. From this data set, students display the information, describe patterns that emerge, and make inferences from the patterns.

Robert Davis (1992) states that it is better to begin with problems, allow students to develop methods for solving them, and recognize that what students take away from this experience is what they have learned. Such learning is likely to be deep and lasting. Exploring problems within a collaborative group is one way to ensure deep learning. It is more appropriate to engage students in solving problems because it is only through problem solving that the concepts and procedures develop together and remain connected in a natural and productive way (Hiebert et al. 1997).

Writing Mathematics Tasks

Not all teachers have access to rich textbooks that align with state or district standards. Some teachers will need to write their own meaningful mathematics tasks. When creating mathematics tasks to use with students, it is important to keep the following in mind:

- Decide on the problem-solving strategy to be used before writing the problem.

- Use appropriate vocabulary for students.

- Keep the sentences simple.

- Allow for multiple entry points and different ways to solve the problem.

- Work the problem at least twice before giving it to students.

- Have another teacher (or a PLC team) work the problem to check for clarity, vocabulary/word choice, and correct mathematics.

Developing a Mathematics Curriculum Timeline

A mathematics timeline is necessary for teachers to see what content needs to be taught and how much time a teacher should spend on a concept. This is the big picture of what the teacher will be teaching throughout the year. The teacher starts with the first month of teaching and writes down what students will learn in conjunction with the standards. It is important to identify the specific standards that will be taught to ensure all standards are being met throughout the year. Then, the teacher identifies when the standards are taught so adequate attention is given to all standards.

The timeline is intended to be an overarching guide to the school year. Instructional decisions should be based on pre-assessments to find out what mathematical content knowledge students already have and what knowledge might be missing. Based on student results, the teacher can make curriculum decisions and focus lessons that correlate with the items for which students did not demonstrate mastery. Then the teacher can complete a more accurate mathematics curriculum timeline. The purpose of the timeline is to reveal the big picture, but the reality is that the teacher will need to make adjustments along the way:

1. Always start by plotting out the mathematical standards.

2. Write learning targets and criteria for success for the standards.

3. Choose the materials, textbooks, lessons, manipulatives, practice games, activities, and resources that will best address the standards.

4. Perform frequent formative and summative assessments before, during, and after units of study to help decide how much time is needed for each concept covered.

5. If applicable, use midyear benchmark/quarter tests as guides for when to introduce concepts during the year.

Figure 3.2 shows a sample mathematics curriculum timeline. To fill out the timeline, teachers follow these steps:

1. Write the day of the week or date in the first column.

2. In the second column, record the content standard and learning targets being addressed.

3. Write the name of the lesson in the third column.

4. Write the mathematics resource or program title in the fourth column.

5. In the fifth column, write which portions of the lesson will be tested and if it will be a formative or summative assessment.

6. In the last column, write any suggested adaptations or notes that teachers may need for each lesson.

Teachers can use this timeline as they complete units of instruction or cover particular standards.

Figure 3.2 Mathematics Curriculum Timeline

Course Unit: _____

Times and Days of Instruction: _____

Date(s)	Content Standard(s), Learning Target(s), and Criteria for Success	Lesson Title(s)	Mathematics Resource	Formative/ Summative Assessments Used	Adaptations or Notes

The use of timelines reminds teachers to find the necessary resources to cover the required content standards. This can involve choosing the appropriate lessons from the adopted mathematics textbook. As a general rule, following a textbook from beginning to end is not the best use of instructional time. Often textbooks have lessons that are not required for given grade levels in a mathematics course. Sometimes textbook lessons are redundant or unnecessary based on student pre-assessments. Therefore, the timeline-planning process requires that teachers find supplementary mathematics resources in order to teach all of the standards.

Sometimes teachers need to plan further practice games or manipulative activities to give students the required practice of a concept. Timelines are also useful for allowing flexibility in meeting student needs for enough time to learn a concept before moving on to a new concept. Teachers can use these resources, along with district or school-site pacing charts, to record progress toward teaching to the content standards. Teachers can also use timelines to demonstrate to administrators, parents, and other teachers that they are teaching the required standards.

Many standards will overlap when teaching concepts, and there are usually critical areas at each grade level that are emphasized. The critical areas are designed to bring focus to the standards at each grade by describing the big ideas that educators can use to build their curriculum and guide instruction (NGA Center and CCSSO 2010). Figure 3.3 shows what these critical areas look like by grade in the CCSSM.

Figure 3.3 CCSSM Critical Areas

Grade	Critical Areas
K	In Kindergarten, instructional time should focus on two critical areas: (1) representing, relating, and operating on whole numbers, initially with sets of objects; (2) describing shapes and space. More learning time in Kindergarten should be devoted to numbers than to other topics.
1	In Grade 1, instructional time should focus on four critical areas: (1) developing understanding of addition, subtraction, and strategies for addition and subtraction within 20; (2) developing understanding of whole number relationships and place value, including grouping in tens and ones; (3) developing understanding of linear measurement and measuring lengths as iterating length units; and (4) reasoning about attributes of, and composing and decomposing, geometric shapes.
2	In Grade 2, instructional time should focus on four critical areas: (1) extending understanding of base-ten notation; (2) building fluency with addition and subtraction; (3) using standard units of measure; and (4) describing and analyzing shapes.
3	In Grade 3, instructional time should focus on four critical areas: (1) developing understanding of multiplication and division and strategies for multiplication and division within 100; (2) developing understanding of fractions, especially unit fractions (fractions with a numerator of 1); (3) developing understanding of the structure of rectangular arrays and of area; and (4) describing and analyzing two-dimensional shapes.

Grade	Critical Areas
4	In Grade 4, instructional time should focus on three critical areas: (1) developing understanding and fluency with multidigit multiplication, and developing understanding of dividing to find quotients involving multidigit dividends; (2) developing an understanding of fraction equivalence, addition and subtraction of fractions with like denominators, and multiplication of fractions by whole numbers; and (3) understanding that geometric figures can be analyzed and classified based on their properties, such as having parallel sides, perpendicular sides, particular angle measures, and symmetry.
5	In Grade 5, instructional time should focus on three critical areas: (1) developing fluency with addition and subtraction of fractions, and developing understanding of the multiplication of fractions and of division of fractions in limited cases (unit fractions divided by whole numbers and whole numbers divided by unit fractions); (2) extending division to two-digit divisors, integrating decimal fractions into the place value system and developing understanding of operations with decimals to hundredths, and developing fluency with whole number and decimal operations; and (3) developing understanding of volume.
6	In Grade 6, instructional time should focus on four critical areas: (1) connecting ratio and rate to whole number multiplication and division and using concepts of ratio and rate to solve problems; (2) completing understanding of division of fractions and extending the notion of numbers to the system of rational numbers, which includes negative numbers; (3) writing, interpreting, and using expressions and equations; and (4) developing understanding of statistical thinking.
7	In Grade 7, instructional time should focus on four critical areas: (1) developing understanding of and applying proportional relationships; (2) developing understanding of operations with rational numbers and working with expressions and linear equations; (3) solving problems involving scale drawings and informal geometric constructions, and working with two- and three-dimensional shapes to solve problems involving area, surface area, and volume; and (4) drawing inferences about populations based on samples.
8	In Grade 8, instructional time should focus on three critical areas: (1) formulating and reasoning about expressions and equations, including modeling an association in bivariate data with a linear equation, and solving linear equations and systems of linear equations; (2) grasping the concept of a function and using functions to describe quantitative relationships; and (3) analyzing two- and three-dimensional space and figures using distance, angle, similarity, and congruence, and understanding and applying the Pythagorean theorem.

(NGA Center and CCSSO 2010)

Maximizing Instructional Time

In today's educational environment of high accountability and high stakes, it is the teacher's responsibility to maximize every possible minute available for student learning. Every mathematics teacher has a favorite activity or game he or she likes to teach—mainly because it is really fun— but if the activity or game does not directly relate to the mathematical content standards and objectives, it should not be used. Each minute of classroom instructional time is crucial and must be directly related to helping students learn mathematics in a coherent and focused curriculum.

Maximizing instructional time requires thoughtful planning before each and every day. The following list provides considerations for teachers when preparing lessons:

- The teacher should have the lesson plan ready and the materials prepared before students enter the room, so no time is wasted while students wait for the teacher to gather the day's instructional resources.

- In a mathematics lesson that involves multiple components (pre-assessment, review, instruction, modeling, guided practice, checking for understanding, group work, independent practice, and formative/summative assessment of the lesson objective), the teacher should look over the various components beforehand to see if any additional materials are needed. The teacher should think through each aspect of the lesson and visualize what is needed.

- The teacher should plan for enrichment activities if there is extra time at the end of the class period or for students who finish early. These activities do not need to be more work, but deeper work or problems focused on the same learning goals/objectives for the day's lesson. This could be an ongoing project or activities developed to help further the learning of students who have already met the learning target. For example, instead of students collecting and analyzing data, they could write a letter to their principal trying to convince him or her to change something based on the data collected. The letter must be coherent and supported with data that was collected within the lesson.

- The teacher should have a plan for interventions for students who are below grade level. This could include alternative assignments/activities, shorter assignments/activities, alternate entry points, or extra time. If students are solving integer problems in a seventh-grade class, all rational numbers should be included in the original lesson. However, if a student does not know how to solve problems with decimals and fractions, he or she will likely not pick up the rules for integers, so it would be helpful to begin with only positive and negative whole numbers. Fractions and decimals can come later when the student has a solid understanding of integer work.

- The teacher should solve lesson problems and have the answers accessible before the lesson begins. This will help the teacher quickly assess the students' answers to pinpoint those who still need help. When developing the answer key, the teacher should find multiple paths to the solution set (right and wrong), so he or she can start thinking about the types of questions to ask students when they work through the problems. Then the teacher can write down some pocket questions for the lesson (as described in Chapter 1).

Prepare in Advance

The first way for teachers to achieve the goal of maximizing instructional time is to be well prepared. It is also helpful to be well acquainted with the content standards and the course concepts that need to be mastered by the end of the year. Teachers can talk to fellow educators in the grade ahead to see what skills they feel are most important for students to have mastered for the curriculum they will cover in the following year. Also, they can talk to the teachers who previously taught their students—they can shed light on what students know and understand, and where any gaps may be.

Teachers should spend time creating daily lesson plans so that instructional time is not wasted on preparation or with the implementation of unpracticed procedures. Teachers can work in teams to develop instructional lessons collaboratively. This is also a great opportunity to define what success will look like when students master a standard or instructional goal. Creating a shared understanding of what student success looks like should be the primary goal of a PLC.

Here is a checklist for teachers to use in preparation of each day's lesson:

- ☐ Select the mathematical content standard.
- ☐ Plan the lesson objectives to be covered in one lesson period.
- ☐ Identify and write student-friendly learning targets and the criteria for success.
- ☐ Choose the sequence or structure of the lesson.
- ☐ Plan lesson activities: pre-assessment, review, instruction, modeling, guided practice, checking for understanding, independent practice, cooperative practice activities, application activities, and assessment of the lesson objective(s).
- ☐ Be conscientious of the level of activity (Webb's Depth of Knowledge 1–4) and maintain that focus throughout the lesson.
- ☐ Gather the materials and resources needed for the lesson.

Implement Efficient Transitions

In a typical mathematics lesson, the teacher may direct students through various modes of activities. A lesson may include whole-class discussion, partner work, small-group discussion, cooperative activities, games, skill practice, and independent work. A lesson will include movement from Launch to Explore to Summary. This requires smooth transitions.

During a given lesson, the teacher will want to maximize the instructional time devoted to mathematical concepts and minimize transitions from one lesson activity to the next. Without clear expectations, rules, and procedures, transitions can cause confusion and take focus away from learning. Transition time is also commonly when behavior problems arise. This is especially true if the proper procedures have not been taught and regularly practiced. Every teacher should strive to establish efficient transition procedures. When transition time in a mathematics classroom is limited, time for learning and applying concepts and skills is increased. Taking extra time at the beginning of the year to model and role-play correct and incorrect transitions will save instructional time later in the year. The saying "We move slowly at first, to eventually move fast" applies here, even for older students.

In initial practice sessions, teachers can prepare students in advance to make the transitions as effortless as possible:

- **Clarify expectations.** The teacher can tell students what they will be doing and check for understanding before allowing them to transition to a new activity: "Look for the criteria for success on the board. We will get together in 15 minutes to review the different properties of triangles and quadrilaterals."

- **Rehearse transitions from various types of classroom activities.** Students will need ample opportunities to practice. The teacher can say, "Okay, let's practice how we begin the activity. Group member #1, get the material. Group member #2, make sure names and activities are labeled, claim it and name it. Group member #3, make sure everyone is ready to work. Okay, let's practice that again!" These transitions have to be practiced, as routines cannot be established without repetition.

- **Choose an auditory or visual signal to help students know when to change activities.** When the teacher uses a signal for transitions, students need time to think before they are required to act. The teacher gives the signal, which instructs students to be silent and informs them of the expectations. Then the teacher gives the signal again to give students permission to move to the next activity. It is important to be conscious of the students' age group when using such signals—for example, squeaky toys might work well with younger students but not with older students.

Possible Signals to Use During Transitions

- Lights turned on and off quickly
- Music
- Classroom chant
- Counting backward
- Raised hand

- Wind chimes
- Thumbs up
- Train whistle
- Rain stick
- Bell or gong
- Call and response

Use Additional Instructional Time

Sometimes a lesson goes faster than anticipated. Teachers should always have a few extra activities planned to make the most of additional instructional time in the classroom. Here are a few suggestions:

- **Explain the concept/summarize the lesson:** Students work with partners to explain the day's concept to one another. This activity gives additional practice and allows students the chance to rehearse the academic vocabulary essential to the concept. Most students better understand a concept once they have had to explain it to someone else; doing so is beneficial for English language learners and struggling students.

- **Personal agendas:** Each student has a chart with a list of activities that are appropriate for him or her to complete independently. These activities can be based on skills needed for remediation or skills for acceleration, depending on the student. When a student has completed an independent activity on his or her personal agenda, that student must obtain the teacher's initials next to the activity's name and description on the chart in order to move to a new activity.

- **Solve to earn a pass:** Each student solves a simple problem orally, or gives an explanation or definition as a "pass" to get into line to exit class, earn a treat, or just celebrate with a high-five.

- **Numbered heads together:** Students quickly form "numbered heads together" groups by getting into groups of four and numbering off one to four. They work together to solve their unique problem given by the teacher that may be review or preview material. They must make sure everyone understands the answer, because the teacher then rolls a die or spins a spinner to determine which number of student from each group will give the group's answer.

- **Problem-solving journal:** Students write about their understanding of the day's learning—for example, vocabulary, concepts, or procedures—in a journal.

- **Exit ticket:** Students respond to a brief prompt or question provided by the teacher. They must submit their response to the teacher in order to leave class or transition to the next activity.

Conclusion

When planning for instruction, teachers need to take time to do the following:

- Unpack standards to identify what students should know and understand in a mathematics classroom
- Identify the essential learning
- Plan when and where the standards are taught
- See the "big picture"
- Plan for the details that help a classroom run smoothly

Effective planning takes time and practice. The more a teacher plans for instruction, the more opportunities students have to learn.

Reflection

1. Locate the state or district mathematics standards for your grade level. First, unpack the standards by yourself. Then compare your thoughts with a content partner to see if you can reach a consensus.

2. Using the unpacked standards, write learning targets and identify the criteria for success.

3. Choose an interactive activity you have used in the past or create a new activity for a concept you are teaching this year. Make a list of the procedural steps, materials, and transition techniques you will use to have the activity run smoothly.

4. Reflect upon a favorite activity that may or may not align with the standards.

Chapter 4

Implementing Mathematical Practice and Process Standards

"Let's face it; by and large math is not easy, but that's what makes it so rewarding when you conquer a problem, and reach new heights of understanding." (Danica McKellar 2014)

Mathematical practice and process standards describe what students need to do and know to master essential mathematics skills and concepts. The Common Core State Standards for Mathematics (CCSSM) Practice Standards, for example, were built on "processes and proficiencies" from the National Council of Teachers of Mathematics (NCTM 2000) process standards of problem solving, reasoning and proof, communication, representation, and connections, along with the strands of mathematical proficiency specified in the National Research Council's report (2001) *Adding It Up: Helping Children Learn Mathematics* on adaptive reasoning, strategic competence, conceptual understanding, procedural fluency, and productive disposition.

Mathematical practice and process standards are generally introduced in kindergarten and emphasized throughout students' K–12 education. These standards should not be based on specific content, but more broadly on how students will learn the content. Students will continually refine their work as mathematicians, and the standards are the vehicle through which students will continue to learn.

Teachers need to be explicit about which state or district practice and process standards are being emphasized in each lesson. Most of these standards will intertwine throughout the lesson, but teachers should emphasize or focus on one or two overarching practice and process standards. When planning a lesson, for example, a teacher implementing the CCSSM would use the chart in Figure 4.1 to identify student language when focusing on one of the eight Standards for Mathematical Practice and highlight teacher questions to emphasize the standard in the lesson. The teacher would write down these questions along with pocket questions for quick access. Students can pick one or two sentence stems daily to use when working with their cooperative group until they naturally develop this language.

Figure 4.1 Questions Aligned to Practice Standards

Practice Standard	Student Language	Teacher Questions
1. Make sense of problems and persevere in solving them.	What are we supposed to do? The problem tells us.... Let's make a plan. How can we monitor our progress? Does this make sense? Let's check our answers.	What is the problem asking? What do you need to find? Have you solved problems like this before? How could you start the problem? What manipulatives/tools might help you? Does your plan make sense? How can you check your answer?
2. Reason abstractly and quantitatively.	What do we know? Does this make sense? How can we label this?	How can you represent the problem with symbols or numbers? What do the numbers or variables refer to?
3. Construct viable arguments and critique the reasoning of others.	How did you get your answer? I think _____ because _____. Why does that work? Explain to me your thinking. I disagree, because _____. Show me how you got your answer.	Compare your answer to your partner's answer. How are they the same/different? What does your answer mean? Can you prove your solution to me? Can you find a counterexample? What questions do you have for your partner about this problem?

Practice Standard	Student Language	Teacher Questions
4. Model with mathematics.	What formula will help us? How can we apply what we know? What do we know? Does this make sense?	What connections do you see? How can you write this using words or symbols? Explain the result to me. Is this true for all cases? Can you predict the next one? What assumptions are you making?
5. Use appropriate tools strategically.	What tools could we use? • Calculator • Computer • Model • Paper/pencil • Protractor • Ruler • Spreadsheet	What tool/manipulative could help you with your thinking? Which tool will help you best solve this problem?
6. Attend to precision.	Are we all discussing this problem? How did you calculate that answer? Can we explain the solution? Should we round?	Explain your thinking. What is the problem asking? Does your answer make sense? How do you know your solution is correct? Does your group agree? Why or why not?
7. Look for and make use of structure.	What patterns do we notice? How can we relate this problem to the real world?	What did you notice when you were working through this problem? Why does this happen? What patterns do you see? Will this work all of the time? Why is this important to the problem?
8. Look for and express repeated reasoning.	Does this always happen? How can we relate this problem to everyday life?	Can you make a rule or generalization? Can you think of a shortcut? How could this problem help in solving another problem?

Processes to Enhance Student Interaction

In most professions, people need to interact and work collaboratively to solve problems. This should be no different in a mathematics classroom. A teacher can create structures so students work cooperatively and support one another as they strive toward mastery of learning. Twenty-first-century skills and Mathematical Habits of Mind/Interaction are structures designed to improve mathematical dialogue and thinking.

Twenty-First-Century Skills

The standards push for productive discourse through similar learning skills, and the Partnership for 21st Century Skills (P21) also emphasizes college and career readiness for every student. To succeed in the twenty-first century, all students will need to perform to high standards and acquire mastery of rigorous core subject material (P21 2011). Teachers can focus on the following twenty-first-century learning skills to help students productively work together to solve meaningful mathematics problems:

- Creativity and innovation
- Critical thinking and problem solving
- Communication and collaboration

Figure 4.2 shows what students need for the twenty-first century and how to support those needs.

Figure 4.2 Twenty-First-Century Student Outcomes and Support Systems

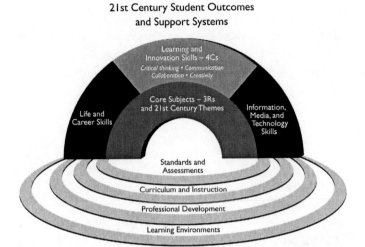

(Printed with permission from P21 2011)

When students are working together, twenty-first-century skills should be highlighted daily. Teachers need to be explicit when teaching students these skills and positively reinforce behaviors or actions when students demonstrate them. Not all students can innately use these skills. For students to know what creativity or true collaboration looks like, it is always helpful to see it in action. To this end, at the beginning of the school year, teachers can directly model some or all of these skills in the following ways:

- Model a think–aloud.

- Model specific thinking strategies.

- Demonstrate how to use manipulatives.

- Analyze data and make interpretations.

- Reflect about the math.

- Demonstrate curiosity.

- Show different ways to think about a problem.

- Role-play a productive mathematical conversation.

- Explain clear thinking.

As students become more comfortable working collaboratively in groups, teachers can have them demonstrate their thinking, working, critiquing, and explaining to the class. This gradual release of responsibility (Pearson and Gallagher 1983) will help students become more confident when thinking and talking about mathematics. By having a model to follow, students will be more likely to leave their comfort zone and take these conversational risks. That is why it is really important for students to be taught how to have mathematical conversations. There is a distinct difference between social interactions and mathematical discourse, and teachers need to be clear about which conversations are appropriate in a mathematics classroom.

P21 (2011) maintains, "As the standards document states, grade-level indicators that begin with the word 'understand' are intended to be good opportunities to connect the practices to content." This is a helpful way to begin thinking about the connections between P21 and the standards. Figure 4.3 shows an example of the connections between P21 and the CCSSM.

Figure 4.3 Connections Between P21 and the CCSSM

P21 Skill	Mathematics Practices
Critical Thinking and Problem Solving	Make sense of problems and persevere in solving them. Reason abstractly and quantitatively. Model with mathematics. Look for and make use of structure.
Communication	Construct viable arguments and critique the reasoning of others. Attend to precision.
Information Literacy	Look for and express regularity in repeated reasoning.
ICT Literacy	Use appropriate tools strategically.

(Adapted from P21 2011)

Mathematical Habits of Mind/Interaction

Teachers Development Group (2011) uses research to provide guidance on the teaching and learning practices that help students learn mathematics. Using Art Costa's work with Habits of Mind, they developed the ideas of Mathematical Habits of Mind and Mathematical Habits of Interaction shown in Figure 4.4. Study this chart to improve how students interact with mathematics and each other.

Figure 4.4 Mathematical Habits of Mind and Interaction

Mathematical Habits of Mind	Mathematical Habits of Interaction
Reason or make sense of a problem	Honor private think time
Justify ideas or answers	Listen to understand
Make conjectures	Ask genuine questions
Use mathematical representations	Analyze thinking
Think about their thinking	Respect others
Make connections	Critique strategies
Use mistakes to start new learning	Persevere
Question their thinking	N/A

(Teachers Development Group 2011)

The Mathematical Habits of Mind/Interaction support the mathematical practice and process standards in a classroom. Teaching students what each habit looks like or sounds like is one step toward getting students to work collaboratively in a group and to grow as mathematicians. The following problem offers an example:

I have a farm and on my farm I have chickens and cows. There are a total of 52 legs and 19 heads. How many chickens and cows do I have?

As with all new skills, the teacher should slowly introduce the habits. The teacher could focus on "Use mathematical representations" and "Listen to understand" as the Mathematical Habits of Mind/Interaction being emphasized. The teacher should honor private think time or private ink time (students can write as they think), and ask students what that would look like or sound like if they were truly listening to others. Then the teacher could fill out a T-chart (see Figure 4.5) with student responses and refer to it often during the class period. The teacher could focus on and point out students who are demonstrating "Listen to understand" and question student language that does not fit on the chart. The teacher could then use this same technique for "Use mathematical representations."

Figure 4.5 T-chart for "Listen to Understand"

Looks Like (Student Actions)	Sounds Like (Student Speech)
All students facing the speaker Good eye contact Listening with both ears No one is talking but the speaker	"What would you do first, count the heads of the chickens and cows or the legs?" "How did you get so many heads? Will it work that way?" "Can you please tell me your thinking?" "So, what you did was add the heads and legs together?" "I heard you say [a student paraphrases another student]...."

It is useful to have Mathematical Habits of Mind/Interaction posters hanging around the classroom throughout the year. Focus on one or two habits a day until they become routine, and then move on to a different Mathematical Habit of Mind/Interaction until students are familiar with all of the habits. The ultimate goal of the habits is for them to be a natural occurrence in any mathematics classroom. More emphasis at the beginning of the year will lead to full implementation later.

The mathematical practice standards, twenty-first-century skills, and Mathematical Habits of Mind/Interaction are all interrelated. It is up to teachers to make the connections explicit for students. To do so in the classroom, the teacher can think of a funnel, as shown in Figure 4.6. In order to achieve authentic student learning, the practice standards provide the overarching goal. However, if the teacher focuses only on these practices, he or she will not get to the practical details that must happen in a mathematics classroom. This is where the twenty-first-century skills and the Mathematical Habits of Mind/Interaction give the teacher tools to actualize the mathematical practice standards in the classroom. This funnel is a tangible way to illustrate the interrelatedness of these three processes.

Figure 4.6 The Funnel of Student Learning

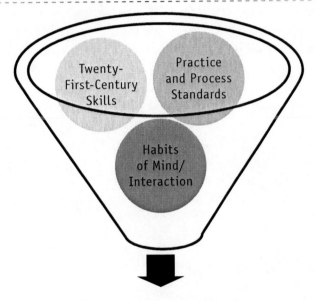

Community of Discourse

Just because teachers put students in groups does not mean students know how to interact with each other appropriately in a group setting. A number of factors prevent students from collaborating, but teachers need to emphasize that the learning environment calls for students to talk with each other daily. Teachers also need to be careful not to get caught up in personal discussions and small talk—classroom time is precious, so it is essential to stay focused on the learning at hand. Some students are quite talented at getting teachers off topic. In this case, the teacher can remind such students that there is a time and a place to have these discussions, but the teacher has only 45, 50, 60, or 90 minutes to help students learn as much mathematics as possible.

Teachers focusing on the "right" conversations will strengthen mathematical comprehension. Most of the mathematical practices and process standards and twenty-first-century skills relate to higher-order thinking skills that will be developed only when students have academic conversations. It will be hard for students to construct viable arguments and critique the reasoning of others, for example, when they have not had the opportunity to listen to or defend other students' mathematical thinking and reasoning. If students are taught only to be the consumers of information whose sole purpose in school is to raise their tests scores, then they are less likely to be successful in high-level courses and jobs in the future (Zwiers and Crawford 2011).

The sections that follow cover the importance of mathematical conversations that should be happening daily in classrooms. Student-centered learning is social, and providing opportunities for students to make sense of the mathematics with others is one step toward deep understanding. However, not all students walk into the classroom ready to have mathematical conversations. The teacher can provide nonacademic structures for students to learn how to have deep, meaningful conversations.

Learning Environment

Students will have a better chance of learning and retaining the mathematics if they are given many opportunities to discuss the content at hand. Students need to reason, clarify, and justify the learning. It is when students start thinking about their own thinking that they begin to refine their mathematical beliefs: "I just realized that .5 always equals ½." Also, when students start to teach others, they tend to gain a deeper understanding of the content: "I realize working in a team isn't always about getting the right answer, but making sure we can explain how we got our answer." In order to learn a subject as difficult as math, students need a learning environment where their needs will be met.

Bob Sullo (2009) identifies three basic needs for students to be motivated to learn:

- To connect and belong
- To be free and autonomous
- To enjoy themselves and have fun

Providing an environment where students are able to have academic conversations allows them a chance to participate and meet the needs of connection and belonging. By giving students the trust that they can and will have discourse in the classroom, the teacher is meeting the need for freedom. When students are engaged in mathematical discourse in a safe setting, they welcome the autonomy to work and defend their thinking. They begin to have fun when they discover that they are successful. Learning can be a joyous occasion. Take time to celebrate any or all victories when students make sense of the mathematics. It can be as simple as a class cheer or putting student names on an "ah-ha" board.

Body Language

One way to introduce mathematics talk is to discuss body language. Ask students what actions or manners they attribute to positive body language. Responses might include the following:

- Maintain eye contact
- Face each other
- Smile
- Display an open stance or lean in as opposed to crossed arms
- Nod to show you are listening

Ultimately, authentic mathematical conversations are what all mathematics teachers want to have in their classrooms, but this is not always the case. Once students understand the power of effective body language, the teacher can then provide structures for students so they have productive mathematical conversations.

Structured Talks

During the first days of school, a teacher can structure casual "talks" or icebreakers as an informal way for students to start communicating with one another. The teacher can have students play human bingo, do a summer vacation scavenger hunt, or participate in some of the other talks in the following list. The teacher should let students know that communication is a high priority in a mathematics classroom, and academic conversations will happen daily. This should first be modeled by the teacher and a student or by the teacher and another adult to help demonstrate to students what mastery looks like. Nonacademic talk could include the following:

- Tell me about your pet, family, vacation, etc.

- What do you enjoy most about school lunch? Least?

- What do you do for fun? What makes it fun?

- What are some characteristics of a good person?

- Tell me a funny story about you, a family member, a friend, etc.

- If you could travel anywhere in the world, where would you go and why?

During these conversations, the teacher can model open body language, how to pause and then paraphrase what the other speaker is saying, and how to ask questions that invite the other speaker to continue speaking. After the conversation, the teacher should ask students what they noticed, write their observations on a board/paper so they are visible, and refer back to them often. As students get used to having conversations with each other, the teacher can have them reflect on their skills and ask if there are more behaviors or actions that could be added to the list.

Mathematical conversations will occur in pairs, in small groups, or among the whole class. They can also happen orally or in writing. Students need to experience the different ways they will be working and conversing with each other because it might change daily. It is helpful if teachers can articulate how students can share their mathematical thinking. Figure 4.7 shows a few ways to structure mathematical conversations in pairs.

Figure 4.7 Conversation Strategies

Conversation Strategy	Instructions
Think-Pair-Share	The teacher asks a question and gives time for student 1 to think of an answer. Student 1 shares his or her thinking with student 2 for a given amount of time. Student 2 then paraphrases the response for accuracy. The teacher asks for student 2 to share the response with the class.
I think...	The teacher asks a question and gives time for partners to think. Student 1 responds, "I think_____, because _____." Student 2 listens silently without responding. After the teacher says "time," student 2 says, "I think_____, because _____." Ask students if they agree or disagree with their partner using the same model: "I agree because _____" or "I disagree because _____."
Paraphrase	The teacher asks a question and gives time for partners to think. Student 1 shares his or her thinking with student 2. Student 2 paraphrases what student 1 said, using phrases such as • So, you are saying... • Let me see if I heard you right... • In other words... • So, you think... Students then switch roles.
Which side?	The teacher shows two methods for a solution. Students decide which method they prefer. They then find a partner who believes the opposite. Student 1 has one minute to defend the method he or she chose. Student 2 listens without interrupting. Student 2 then has one minute to defend the method he or she chose. Student 1 listens without interrupting. The teacher asks students if they would switch methods after listening to their partner.

Conversation Strategy	Instructions
What I think you are thinking is...	The teacher has students show their written work to a partner. Both students take time to read through the work. When the teacher gives the verbal cue, student 1 states, "What I think you are thinking is..." Student 2 will agree or disagree and state why.
	Students then switch roles.

Grouping Strategies for the Classroom

No single method is best for arranging students in groups. Many teachers find that when students get to pick their group members, they are more comfortable talking with one another, but they might tend to get off task more easily. When teachers choose groups at random—for example, by using matching cards from a deck—some students will be happy, while others will be upset. When teachers pick groups, this can be a more balanced approach when it comes to students' abilities. It is easier to match high-performing students with high/middle-performing students and middle/low-performing students with low-performing students.

There is no timeline for how long groups should work together, but it is important for students to be really comfortable with one another before switching groups. Some teachers group students on a month-by-month or unit-by-unit basis. In a mathematics classroom, it is important for students to work with others at all learning levels.

Teachers can try out some of the different grouping techniques discussed in more detail in the sections that follow and do what feels most natural. Options include the following:

- Independent learning
- Paired learning
- Cooperative learning
- Whole-group learning

Independent Learning

Having students work cooperatively with others is instructionally sound, but it should not be viewed as the only way to effectively teach students. They also need opportunities to explore on their own and to examine, question, and hypothesize about their own learning. Independent learning develops students' confidence and natural curiosity. Students will have greater success with independent work if it directly relates to the skills they are working on during guided-practice activities. When students are engaged in independent work, the primary goal of the teacher is to monitor their processes. As students independently work, the teacher should closely monitor all students. The work during independent practice is usually a good indicator of how well students know the mathematical content. This is also a good time for teachers to give lower-ability students one-on-one or small-group instruction.

Paired Learning

Working with a partner allows students more time to process new content and practice applying new knowledge in a low-stress atmosphere. Paired learning is a useful grouping strategy in a mathematics classroom in many instances:

- Often when manipulatives are being used, two students can sit side by side and share materials so that each has an opportunity to work with the materials.

- When students need to rehearse information about concepts they have learned or solve a problem together, they can discuss the steps necessary before sharing their work with the class.

- English language learners frequently need to practice using the necessary vocabulary for each mathematical concept. They need time to say the vocabulary words, use them in context, and get immediate feedback about what they said.

Cooperative Learning

At times it is appropriate for teachers to organize students into small groups to explore a mathematical concept. Within each group, students are given instructions for completing a task. To make cooperative grouping effective, teachers need to consider students' achievement levels. Students can be organized into heterogeneous groups in which students' ability levels are varied, homogeneous groups in which students have the same ability levels, or flexogeneous groups that combine heterogeneous and homogeneous groups within one lesson. Within a mathematics classroom, all three grouping techniques can be used. Teachers should evaluate lessons to determine which technique will work best for each individual lesson.

Teachers can take certain steps to ensure participation in cooperative learning groups. When groups are larger than two, it is sometimes helpful to assign group roles. Please note that not every group needs defined roles; some groups work cooperatively from the start. If some groups are slow to start working, or are new and uncomfortable with each other, it can be helpful to implement team roles. Team roles should switch daily so every student can try a new leadership role. The ultimate goal for collaborative groups is for students to converse naturally, but when that does not happen, team roles can help break down barriers.

Figure 4.8 shows many different roles a teacher could use for groups. Which type of group the teacher uses depends on how many students are in a group. Students should practice the responsibilities for each role. It often works best if the teacher picks the role until the responsibility is routine. It is fine to combine roles.

Figure 4.8 Cooperative Learning Group Roles

Role	Responsibilities
Facilitator	• Makes sure everyone is on task • Makes sure everyone is participating equally • Makes sure all voices are heard • Questions all group members • Paraphrases what other group members say
Reporter	• Makes sure all group members can articulate the method and solution to the problem • Makes sure all group members understand the mathematical concept
Reader	• Reads the problem • Clarifies questions about the problem
Coach	• Provides positive support for the group • Makes sure everyone works together
Timekeeper	• Watches the time • Keeps everyone on task
Recorder	• Makes sure everyone has methods and solutions to the problems • Makes sure all papers are titled appropriately (name, date, assignment, class period, etc.)
Materials manager/ supply sergeant	• Gathers manipulatives/tools as needed • Makes sure everyone has a book, paper, a pencil, etc.
Idea person	• Asks questions such as • What if we tried...? • I wonder if we could...? • How about we...?
Challenger or devil's advocate	• Asks questions such as • Is this right? • Why do you think that? • Are there other methods? • What other ways could we do this?

Role	Responsibilities

Checker	• Asks questions such as
	  • Is the solution right?
	  • Are there multiple solutions?
	  • Can everyone in the group articulate the method and solution used?

Equity monitor	• Asks questions such as
	  • Is every group member participating?
	  • Are all voices being heard?
	  • Can we take a minute to think about what you said before we respond?
	  • Are we all on the same problem?
	  • _____, what are you thinking?

One of the best ways for students to observe and reflect on student discourse is to host a Fishbowl, also known as a Socratic Seminar. In a Fishbowl, the teacher chooses a deep mathematical task that may have multiple entry points. Three or four students are in the middle (as fish) and work on the task. While the fish are engaged in the task, the rest of the class is silently observing the group's interaction skills. The observers take notes on the fish. It is helpful to give prompts to help students get started, such as "I notice..." and "I wonder..."

Additional strategies for the Fishbowl include the following:

• Each fish has a partner or "helper" if he or she gets stuck. The helper must remain silent unless directed to speak by the teacher.

• The teacher calls a huddle for the helpers to brainstorm ideas or strategies they could share with their fish.

• The teacher gives observers specific behaviors to watch for, such as listening skills, body language, or types of questions being asked. The observers record these behaviors and share their observations with the fish.

It is good to have student observers reflect on what they saw, heard, or noticed. These observers should be reminded that the comments they provide need to be helpful and not hurtful. After the Fishbowl activity, it is beneficial for students to journal. Taking time to write and reflect helps students think about their own actions as group members and consider ways they can improve. Also, the teacher needs to take time to respond to students' writing to show that he or she cares about how students interact with one another.

Whole-Group Learning

Whole-group learning can take place at many different times. For example, if the teacher notices ongoing confusion with a mathematics concept, he or she can "catch" the class and ask a clarifying question or teach a part of the concept. Another time for whole-group learning is at the end of the lesson. Students need to make sense of the mathematics they were participating in and show and tell what they learned to the rest of the class. The teacher acts as a facilitator and structures the students' mathematics talk. It is during this time that students question or make comments on each other's thinking.

When introducing a new routine, teachers can use whole-group time to practice and discover how this will help make a safe classroom where authentic mathematical conversations can take place. When students feel trust in a whole-group setting, they are ready to take risks, respect others, and learn.

Student Feedback and Reflection

For student discourse to happen daily, the teacher and students must be equally committed to its success. The teacher must show up with prepared lessons that are standards driven. The method by which the teacher delivers the lesson may vary, but it is crucial to plan for student talk and for students to take an active, meaning-making approach to learning.

It takes time and effort for students to become comfortable with accountable talk, but it needs to be the expectation in a mathematics classroom. It is the teacher's responsibility to give effective feedback on how students are performing in small groups and individually. The teacher's role in fostering individual and group accountability is to create assessment systems that encourage feedback among the members of the group and between the teacher and each student (Frey, Fisher, and Everlove 2009).

One way a teacher can provide feedback is through observation and scripting. The teacher can take the time to watch groups work and write down effective language and ineffective language the group is using. Then the teacher shows the group his or her observations and asks them what they notice. Do they agree or disagree with the observations? Could they add any additional information (positive and negative)?

Students also need to reflect on their actions as productive group members. At the start of class, the teacher can ask students to set a personal group interaction goal and then have students rate themselves at the end of class. The teacher should give students time to write a reflection on their rating with evidence supporting the score.

The use of a rubric can also help students monitor their progress as productive group members. The teacher can have each group pick five essential group behaviors (or use all of them) and rate themselves on the following scale:

- 3 = Effective group member
- 2 = Somewhat effective group member
- 1 = Not an effective group member

The teacher asks students to give evidence to support their scores. Once students are used to rating themselves, they can rate each other, supplying observations to support their choices. Figure 4.9 shows a sample rubric.

Figure 4.9 Productive Group Behaviors Rubric

Group Members: _____

Name: _____

Group Behavior	3, 2, 1 – Provide Evidence or Describe Why
Contributing	
Sharing	
Helping	
Listening	
Asking questions	
Justifying why	
Being ready	
Working equally	
Thinking hard	
Finishing work	
Final comments:	

It is beneficial for classes to develop their own set of productive group behaviors. If students have a say in what is important to them and what they value from others, they will have more buy-in and will likely take the rubric more seriously. This approach is much more effective than a teacher-mandated rubric that students must adapt to.

Conclusion

Standards do not only allow for mathematical content knowledge, but also for explaining, reasoning, critiquing, and justifying why mathematics works. As teachers plan their lessons, they should provide opportunities for students to think about and discuss mathematics. Learning is social and mathematical conversations will help deepen students' understanding.

Reflection

1. How are the mathematical practice standards, twenty-first-century skills, and Mathematical Habits of Mind/Interaction related? Why are they important? Why might using all three be effective in a mathematics classroom?

2. Using a lesson that you have taught or are going to teach, explicitly plan for implementing mathematical practice and process standards, twenty-first-century skills, or Mathematical Habits of Mind/Interaction.

3. Why is it important to develop a community of discourse within your classroom? What are some steps you can take to make that happen?

Chapter 5

Building Conceptual Understanding

"The challenging and interesting tasks found in application problems help teachers engage students in learning." (Seeley 2004)

This chapter offers multiple examples to teach students problem-solving strategies. As students work through the different strategies, they are provided with diverse opportunities to make sense and meaning of the mathematical approaches to solving problems.

This chapter also provides teachers with resources for developing the specialized vocabulary necessary for mathematical concept comprehension. Various vocabulary-development activities are available so that students, including English language learners, can truly understand the academic vocabulary that will help them unlock the mathematical concepts.

Students as Problem Solvers

Mathematical reasoning and problem solving are crucial to true conceptual understanding. Students need time to explore rich mathematical tasks to develop knowledge and insights that make sense to them. Drilling students on skills does not foster understanding; students need to make mathematical connections by communicating with others for a deeper purpose. The presentation of an explanation, no matter how brilliantly worded, will not connect mathematical ideas unless students have had ample opportunities to wrestle with examples (Zemelman, Daniels, and Hyde 2005). It is no surprise then, that problem solving is an important practice that requires ongoing focus in the classroom. Most—if not all—important mathematics concepts and procedures can best be taught through problem solving (Van de Walle and Lovin 2006).

While students must be skilled at performing computations required to find mathematical solutions, this is only part of the process. John Van de Walle and Lou Lovin (2006) state that mastery of basic skills can be embedded in a problem-based approach to learning. Before students begin to manipulate the information in a problem, they should understand its meaning and plan a way to solve it. Focusing and refocusing on skills will not help students learn math, but helping students make sense of the mathematics will help develop confident mathematicians.

Students therefore need to learn about problem solving as a process and the strategies they can apply to find solutions (National Research Council 2001). The process of problem solving goes beyond finding simple solutions—it encourages students to reflect on the solutions, make generalizations, and extend problems to include new possibilities for investigation and be flexible in their approach (Woodward, et al. 2012). Once students learn the process of problem solving, they can use mathematical approaches to solve real-life problems (Gojak 2011).

The sections that follow cover the problem-solving process and creative ways for students to explore math.

Steps for Problem-Solving

The following four steps enable students to tackle problems in a meaningful way (Polya 1945):

1. Understand/unpack the problem.

2. Plan a solution.

3. Carry out the solution.

4. Look back and explain.

It is helpful if teachers explicitly demonstrate the steps for problem solving, model them, and finally allow ample opportunities for guided and individual practice as students approach the problem-solving strategies.

Step 1: Understand/Unpack the Problem

A teacher should encourage students to read a given problem carefully a number of times until they fully understand it. As students learn this step and progress toward internalizing it, the teacher will allow time for students to discuss the problem with peers or rewrite the problem in their own words. This is an appropriate time for the teacher to honor private think time or private think/ink time. Students should ask internal questions such as "What is this problem asking me to do?" and "What information is relevant and necessary for solving this problem?" For example:

Sarah was putting together bags of candy for her birthday party. In each bag, she put 4 pieces of taffy and 3 candy necklaces. If Sarah has 15 bags, how many candies did she use in all?

Next, represent this scenario with a numerical expression.

It is important to give students private think time to make sense of the problem: "What are some clues that will help me find the total number of candies?"

Then, have students share their thinking with partners to see if they are on the same track:

Student A: "I think we could add all the candy first and then multiply it by the number of bags."

Student B: "We could also figure out the total number of taffy and necklaces and then add them together."

Students are now ready for Step 2.

Step 2: Plan a Solution

Students should decide how they will solve the problem by thinking about the different strategies they can use. Often students must use multiple methods to solve problems, and sometimes this is difficult for them to understand. Students should be encouraged to take risks in their thinking and methods. When getting the answer becomes the focus of a lesson, rather than doing the thinking to solve the problem, students gravitate to the easiest method available to get the answer (Van de Walle and Lovin 2006). It is important for the teacher to record the strategies students choose and to have students justify their thinking and reasoning behind those strategies. This record will be invaluable during the summary of this lesson.

Using the previous problem, the teacher could encourage students to discover multiple ways to find a solution that inspires students to think creatively.

Student A: "If we add 4 and 3 first, we know how much candy is in each bag, and then we can multiply by the number of bags."

Student B: "First, we could find out how much taffy is in all the bags (4 × 15), and then find the total number of candy necklaces (3 × 15), and add those two solutions together (60 + 45)."

Step 3: Carry Out the Solution

Students should solve the problem by executing their plans. As they attempt different strategies, they should keep track of their thinking to help them make conclusions later on. The "Problem-Solving Strategies" section later in this chapter describes 12 such strategies.

Step 4: Look Back and Explain

Many times, the solution and strategies in one problem can help students know how to solve another problem. Teachers need to teach and model for students the critical process of reflection. Teachers can even solve problems incorrectly in order to go through the reflection process and "catch" mistakes. To facilitate reflection, students can decide if their answers make sense and if they have answered the question that was asked. They should illustrate and document in writing their thinking processes, estimations, and approaches. This gives them time to reflect on their practices and grow as problem solvers. James Hiebert et al. (1997) state that "reflecting and communicating are the processes through which understanding develops." When students have an answer, it is helpful to have them justify their thinking with others.

Using the previous candy example, the teacher can ask students how the two solutions are related. This "catch" can happen during the Explore phase of a lesson or during the Summary phase: "Here is where the numerical expression is uncovered."

$$15(4) + 15(3)$$

$$15(4 + 3)$$

The teacher can have students show their work to the class as they explain their thought process. It is important to find different solution paths for students to share and to strategically pick students who have different pathways. Here, student work complements the structure of the distributive property. An abstract concept becomes concrete, supporting the vertical alignment in future mathematics classrooms. To help students internalize this concept, the teacher should allow students to solve problems given the same parameters.

Creative Ways to Explore Mathematics

Students need ample opportunities to think about mathematical concepts, relate them to their everyday lives, and successfully perform mathematical procedures. Effective practice does not mean students complete the even-numbered problems in class and the odd-numbered problems for homework. This approach does not allow students to adequately explore successful application and mastery of mathematics. The most important thing teachers can do is allot sufficient time for students to interact and learn together. Teachers can structure this time with rich tasks so students have multiple opportunities to interact with others and the mathematics.

The creative activities described in the sections that follow allow students interesting ways to discuss mathematical concepts.

Peer Explanation Activity

After students have been introduced to a new mathematical concept, the teacher divides the class into pairs and gives partners a different problem to solve. Then, the teacher can direct each pair to join another pair and explain their solutions to one another. This activity can help students reinforce procedure and concept understanding.

Whiteboard Collaborative Group Activity

Students work in groups of four, each with a number from one to four. Each group has a small whiteboard and a pen. The teacher displays a problem and directs all students to solve the problem individually on paper. Then, the group members compare their solutions. If someone has a different answer, the group can discuss the differences. Once everyone has agreed on a method to a solution, the teacher chooses a number from one to four. The group member whose number corresponds to the number given by the teacher records the group's method to the solution on the whiteboard. That student displays the answer to the class before receiving a new problem to solve in the same way. If any group has a different method to the answer, the class can discuss the methods together or in their small groups.

Oral Rehearsal Activity

The language of mathematics requires specific formation of English syntax and new terms. Students need opportunities to practice "speaking mathematically." In the typical question/answer classroom dynamic, only the confident students raise their hands to answer questions. Teachers should structure oral practice so that *every* student is justifying his or her thinking and being allowed to share with a partner. This approach supports collaboration in groups so every student is working, thinking, speaking, and justifying, and all students are prepared to respond. The teacher uses various ways to structure these random responses. A few structures include drawing craft sticks with student names, notecards with student names, or dice with corresponding team and individual numbers.

In another oral activity, the teacher can display a problem on the board, ask for a volunteer to read it, and then continue with that *same* example (rather than writing out another one) so that *each* student can choose to read it. An example would be reading a place-value number, such as 23.4567, correctly. In this activity, each student gets the opportunity to orally practice, and all students can participate within their own comfort level. It may seem redundant to have every student repeating the same number, but it is through repetition that learning sticks.

Graphic Organizer Review Activity

This activity allows students multiple opportunities to review and reinforce crucial information. Before a test or at the end of a unit, the teacher divides the class into pairs. The teacher then gives each pair of students matching colored paper. No two pairs can receive the same color. Each pair graphically displays the important information from that unit on their colored paper as a review or study guide. Partners should decide on a way to display the information that makes sense to each student and clearly shows the information the pair wants to include. Partners should be encouraged to write explanations or steps, draw pictures or diagrams, and show sample problems that help them review any mathematical concepts and problems they have been learning. Each member of a pair should have the same information on his or her paper.

When all pairs are finished creating their graphic organizers, students split from their partners to find new partners with different colored papers. They each share their graphic organizers and verbally explain the information they included. Next, they add any new information learned that was not already included on their papers. Then, they each find new partners and repeat the same sharing process.

The teacher can also decide to give a basic outline of the essential learning to students before they begin working. A time limit on this activity can be set: "You have five minutes to share your work with as many different partners as possible." Or the teacher can require each student to share with a set number of other students: "You have two minutes to find a partner and each share your thinking." To summarize this activity, the teacher can ask students to share with their original partner what they learned and then have the whole class share out.

Working as a Collaborative Team Activity

Students can work in pairs or in small groups. Each person on the team needs a different colored pencil. The teacher displays a problem. Students each write one step to the problem and then pass the paper to another group member. While the other students may offer suggestions, help, or advice, they may not take the colored pencil from the student whose turn it is. Students continue to rotate until the problem is solved. When monitoring the work, the teacher can quickly observe who contributed the most to the problem and who still needs additional help. This is an ideal place for the teacher to check for student understanding of a skill such as balancing equations.

Problem-Solving Strategies

The sections that follow contain explanations and examples of 12 problem-solving strategies that teachers can adapt to meet students' needs. Insights into ways a particular strategy can be used and examples are provided to demonstrate the application of the strategy to the solution of the problem. These examples are not appropriate across all grade levels—they are used only to demonstrate the use of the strategy.

Strategy 1: Guess and Check

This strategy allows a student to make an educated guess and check the guess against the parameters of the problem. If the guess is not a correct solution, the student revises the guess and checks until a correct solution is found. Each student begins by finding the important facts in the problem and making a reasonable guess based on the information.

Example: Maria is 5 years older than Saul. The sum of Maria's age and Saul's age is 25. What are their ages?

Saul's Age	Maria's Age (I know she is 5 years older than Saul.)	Saul's Age + Maria's Age (Must = 25)
5	5 + 5 = 10	5 + 10 = 15 Oops, too low.
12	12 + 5 = 17	12 + 17 = 29 Oops, too high.
10	10 + 5 = 15	10 + 15 = 25 Got it!

Solution: Saul is 10 and Maria is 15.

Strategy 2: Create a Table

This strategy helps students organize information so that it can be easily understood and relationships between the sets of numbers become clear. A table makes it easy to see the known and unknown information. The information often shows a pattern or part of a solution, which can then be completed. It also can help reduce the possibility of mistakes or repetitions.

When using tables for the first time, teachers may have to help students decide on the number of columns or rows to fit the variables. First, teachers should determine how to classify and divide the information. Then they establish the number of variables to be included in the table. Next, they decide on the number of rows, columns, and headings. Otherwise, teachers can try letting students explore different numbers of inputs and outputs that they might need to construct a worthwhile table.

Example: There are 18 animals on the farm. Some are chickens and others are cows. There are 70 legs visible. How many chickens and how many cows are there?

Number of Chickens	Number of Cows	Total Number of Animals = 18	Number of Legs = 70
5	13	18	5(2) + 12(4) = 58 Oops, too low.
10	8	18	10(2) + 8(4) = 52 Oops, too low. I am getting further away.
1	17	18	1(2) + 17(4) = 70 Correct!

Solution: Students will need to draw a four-column table. There are 17 cows and 1 chicken.

Often a pattern becomes obvious when creating a table. Students may leave gaps in the table and complete the patterns mentally or follow the pattern in the table to find the information the problem is asking for.

Example: Two people's ages are being compared in this problem: Mrs. Crawford is 32 years old, and her daughter Lisa is 8 years old. How old will Lisa be when she is half as old as her mother?

Lisa	Mrs. Crawford
8	32
9	33
10	34
11	35
12	36
13	37
24	48

Solution: Students will need to draw a two-column table. By leaving gaps and calculating mentally, they can establish that when Lisa is 24 years old, her mother will be 48 years old.

The next example shows how a pattern can be established.

Example: A child is playing a game of basketball by himself in the park. At regular intervals, other groups of children arrive at the park. From each new group, 2 children decide to join the basketball game. The first group has 3 children, the second group has 5 children, and the third group has 7 children. How many groups will have appeared by the time there are 64 children in the park?

Groups	Children	Total
	1	1
1	3	4
2	5	9
3	7	16
4	9	25
5	11	36
6	13	49
7	15	64

Solution: After setting up the table, students will find the pattern. Seven groups will have appeared.

Strategy 3: Act It Out or Use Concrete Materials

This strategy uses objects or materials to represent people or things in the problem, which helps students visualize the problem in a concrete way to find the solution. A variety of objects such as beans, counters, blocks, toys, and erasers can be used to symbolize people or places. These objects can be moved through the steps of the problem. It is important to chart this movement to keep track of the process. Students can also act out the roles of the different participants depending on the situation and the problem.

Example: There are 10 people in a room. If each person shakes everyone's hand once, how many handshakes will take place?

Solution: Have 10 students act out the solution to this problem and keep track of the handshakes. There will be 45 total handshakes.

Strategy 4: Draw a Diagram

This strategy often reveals aspects of the problem that may not be apparent at first. A diagram that uses basic symbols or pictures or a simple line drawing may enable students to visualize the situation more easily and can help them keep track of the stages of a problem in which there are many steps.

Example: How many markers are needed if you place a marker at every 2-meter point on a 10-meter rope?

Solution: In response, students may mentally calculate $10 \div 2 = 5$; 5 markers are needed. However, if students draw the rope and markers, they will see that 6 markers are needed because an additional one is needed for the starting point.

Using a time/distance line to display information helps to show distance or movement from one point to another.

Example: A signpost is placed on the highway. It says that the city is 30 kilometers to the west and the ocean is 65 kilometers to the east. How far are you from the city when you are 17 kilometers from the ocean?

Solution: Students could draw a line and write the distances on it.

30 km + (65 km − 17 km) = 78 km

or

$$(65 \text{ km} + 30 \text{ km}) - 17 \text{ km} = 78 \text{ km}$$

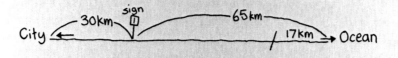

Students need to understand how to plot a course by moving up, down, right, or left on a grid, or use the compass points to direct themselves—north, south, east, west, northeast, southwest, and so on.

Example: Plot four different routes from Byamee to Gumpy without passing through any town twice per route.

Solution:

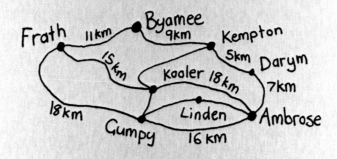

Students will find it helpful to draw diagrams and use symbols to show the relationships among things.

Example: John, Jack, and Fred love animals. John's favorite animals are fish and horses. Jack's favorite animals are horses and rabbits. Fred's favorite animals are fish and rabbits. Which boys have fish as their favorite animals?

Solution:

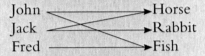

Drawing pictures can help students organize their thoughts and simplify a problem.

Example: Organize the 4 domino pieces in a square shape with each side of the square adding up to a total of 10.

Solution:

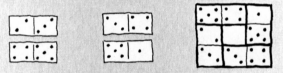

Strategy 5: Create an Organized List

This strategy is used instead of a table when a greater amount of information is available. Students need to follow a procedure or sequence to find the solution to the problem and make sure no information is left out or repeated. Students need to write down the processes they are using to keep their thinking and work organized. Students can work sequentially, using the information from the problem or filling in the gaps of a pattern once it is created. Systematic work is the key to success with this strategy.

Example: Shaun has 3 toy cars that he keeps on his bookcase. One is red, one is blue, and the third is green. He likes to change the order in which he displays them. How many different ways can he do this?

Solution: To solve this problem, place the red car (r) first, and then place the other cars in their 2 possible positions. There are 6 different possibilities.

Strategy 6: Look for a Pattern

This strategy is an extension of drawing a table and creating an organized list. It is often used because mathematical patterns can be found everywhere.

Here are a few ways to check for a pattern:

- Some patterns are additive (2, 4, 6, 8, ...).

- Some patterns are multiplicative (2, 4, 8, 16, 32, ...).

- Some patterns include two operations (6, 9, 8, 11, 10, ...).

Example: When Jim went strawberry picking, 1 out of every 6 strawberries was unripe. How many ripe strawberries were there out of 84?

Solution: The ripe column will increase by multiples of 5, and the unripe column will increase by intervals of 1. So, if a total of 84 strawberries are picked, 70 will be ripe and 14 will be unripe.

Ripe Strawberries	Unripe Strawberries	Total Number of Strawberries
5	1	6
10	2	12
15	3	18
20	4	24
25	5	30
30	6	36
35	7	42
40	8	48
45	9	54
50	10	60
55	11	66
60	12	72
65	13	78
70	14	84

Strategy 7: Create a Tree Diagram or Table

To create a tree diagram, students use branches to represent relationships between different factors in a problem. A tree diagram enables students to visualize the different factors in the problem and ensures that no factors are repeated or missed. First, students need to identify the important factors in the problem and list them. Then, they connect the factors using lines or parentheses. When the problem is complete, students check it to make sure that all factors are properly connected and no information has been left out.

Example: You are at an ice cream shop that offers waffle cones and sugar cones. The shop's specialty ice cream flavors are chocolate, vanilla, and strawberry. How many different combinations of cones and specialty ice cream flavors could you choose from?

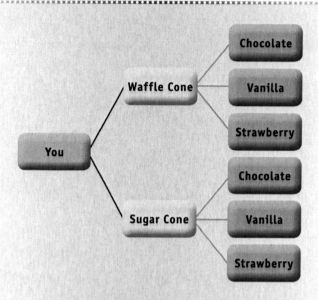

Solution: There are 6 different combinations. The information from the problem can also be organized in a table, which will also show 6 different combinations.

	Chocolate	Vanilla	Strawberry
Waffle cone	WC	WV	WS
Sugar cone	SC	SV	SS

Strategy 8: Work Backward

This strategy is used to solve problems that include a number of linked factors or events, where some of the information has not been provided. The object is to determine the unknown information. The events occur one after the other, and each stage or piece of information is affected by what comes next. To solve the problem, students begin with the ending information and work backward until the problem is solved. It is important to remember that mathematical operations will have to be reversed.

Example: Caleb earned an allowance on Monday. On Tuesday, he spent $5.00 on a used video game. On Wednesday, his sister gave him the $2.50 she owed him. On Thursday, he spent $1.00 on a candy bar. If he has $5.00 left on Friday, how much is his allowance?

Solution: Students can use the following equation to solve the problem:

$$\$5.00 + \$1.00 = \$6.00 - \$2.50 = \$3.50 + \$5.00 = \$8.50$$

Caleb's allowance is $8.50.

Strategy 9: Solve a Simpler Problem

Students can use this strategy to solve a difficult or complicated problem in order to simplify the numbers. Students begin by substituting smaller numbers for larger numbers to make the calculations easier, and then use the same steps to solve the original problem. Alternatively, they can solve a series of simpler problems to see if a pattern emerges, and apply the pattern to the more complicated problem.

Example: You have a 5 × 5 game board. How many total squares are there?

Solution: Students keep track of the number of squares in a table, and start with the smallest grid pattern and work their way to the largest grid.

"I can count 25 squares in the 1 × 1 smallest grid pattern."

"I can count 16 squares in the 2 × 2 grid pattern."

"If I continue this method, I can keep track of the grid pattern and total number of squares that fit the description."

Grid	Number of Squares
1 × 1	25
2 × 2	16
3 × 3	9
4 × 4	4
5 × 5	1
Total number of squares:	55

Strategy 10: Account for All Possibilities

This strategy is used to explore problems that might be answered in a number of ways. Although finding the correct answer is important, teachers should value each student's process for solving the problem and gain insight into the student's developmental understanding. When structuring problems, teachers should use words such as *create*, *make*, *design*, *explore*, and *investigate*. Students should use processes such as labeling counters to visualize the problem, trying different combinations of numbers, and working to find as many solutions as possible.

Example: Investigate which combinations of the digits 3, 4, 6, and 7 will create addition problems that have a sum falling between 100 and 120. For example:

"If I add 3, 4, 6, and 7, I know that I will not get a number between 100 and 120. But I could try combining the numbers and adding 34 and 67, and ah-ha! It's between 100 and 120."

Teachers should encourage students to find multiple combinations.

Solution:

43	34	43	34
+ 76	+ 67	+ 67	+ 76
119	101	110	110

37	36
+ 64	+ 74
101	110

Strategy 11: Analyze and Investigate

When using this strategy, students analyze what is known and what needs to be known. They use the known information to investigate the problem and collect data. Students begin by making an estimate, which is an effective method for gauging the reasonableness of an answer. After estimating, they plan an approach to solve the problem. They determine what is involved in the task and what strategies should be used to gather information. Finally, they decide how to present the information gathered.

Example: Each weekday, cars continually stream past the school. How many weekdays will it take for 1 million cars to pass the school?

One possible solution: Students believe there are two peak traffic times: 8:00–9:00 a.m. and 3:00–4:00 p.m. During the remainder of the day, cars do not pass the school often.

Plan: Survey traffic during the peak hours and at other times during the day. Tally the number of cars that pass during a 10-minute period, and multiply by 6 to get the number of cars per hour.

Peak hours:

8:00–9:00 a.m., 250 cars in a 10-minute period: $250 \times 6 = 1{,}500$ cars/hour

3:00–4:00 p.m., assume the number of cars passing is the same as in the morning: 1,500 cars/hour

Other times:

1:00–2:00 p.m., 10 cars in a 10-minute period: $10 \times 6 = 60$ cars/hour

The period between midnight and 6:00 a.m. would be very quiet, so estimate 5 cars per hour during this time.

The flow of traffic in one 24-hour period can be recorded as follows:

Peak hours = 3,000 cars

Midnight to 6:00 a.m. = 5 cars/hour × 6 hours = 30 cars

All other times = 60 cars/hour × 16 hours = 960 cars

Total cars in one weekday = 3,000 + 30 + 960 = 3,990

1,000,000 divided by 3,990 cars/day = 250.6 weekdays until 1 million cars pass the school

Strategy 12: Use Logical Reasoning

Students can use this strategy when the problem gives information as pieces of a puzzle. Each piece of information is important to solve the problem. Process of elimination is one approach in this strategy, where each piece of information builds to the solution. Drawing a grid to organize given information is also an approach that can be used. Students start by reading each clue carefully and thoroughly. Often students need to deal with the clues in an order different from how they were presented. Then they can take the steps necessary to solve the problem.

Example: Julie, Yushiko, and Sam are each about to eat a sandwich for lunch. On the plate are a tomato sandwich, a honey sandwich, and a peanut butter sandwich. Use the following clues to work out which sandwich belongs to each person:

- Julie's sandwich has salt and pepper on it.
- Sam is allergic to nuts.
- Yushiko dislikes sweet things.

Solution: By drawing a grid, students can use the information they learned in the problem to cross out information and visualize the correct answers.

Sam cannot have the peanut butter sandwich since he is allergic to nuts. Yushiko will not eat the honey sandwich because she does not like sweet things. Julie's sandwich has salt and pepper on it, which means it is not the peanut butter sandwich or the honey sandwich. From this information, students can fill in the grid to show who gets which sandwich.

	Julie	Yushiko	Sam
Tomato sandwich	Y	N	N
Honey sandwich	N	N	Y
Peanut butter sandwich	N	Y	N

Using Mathematical Tools

Research shows that students gain more conceptual understanding and are more successful in demonstrating mastery of concepts when they have had a chance to concretely experience mathematical concepts using mathematical tools (Hiebert et al. 1997; Leinwand 2009; Tapper 2012). James Hiebert et al. (1997) state that tools shape the way we think about problems and influence the methods we develop for solving them. In addition, when students use tools, they perform better academically and have more positive attitudes toward mathematics (Leinenbach and Raymond 1996). Students begin to gain confidence and deeper understanding when they have a model to use and explain their thinking.

Some teachers may think that students will misuse tools, rather than focus on how tools or models help with the lesson concept. Teachers might need to learn how to use different tools effectively in order to make abstract concepts concrete and believe that it will take extra time to prepare, pass out, and collect the tools. It is vital for teachers to use multiple tools to solve problems and understand that students will use tools differently to construct their own meaning. Tools are meant to support learners as they explore and make connections.

As they gain more experience using tools and the tools become part of the everyday routine, students will come to appreciate the practice and application that tools add to their mathematical learning experience.

Tools can refer to any language, material, and/or symbol being used in the classroom as external support for learning (Hiebert and Grouws 2007). It is important for teachers to emphasize the positive impact of models and tools in the learning of mathematics.

Figure 5.1 describes some mathematical tools and their uses.

Figure 5.1 Mathematical Tools and Possible Applications

Tools	Possible Uses in a Mathematics Classroom
2D shapes	• Geometry • Sorting and classifying • Comparing and ordering • Recognizing shapes • Describing • Drawing • Area and perimeter
3D shapes	• Geometry • Sorting and classifying • Comparing and ordering • Recognizing shapes • Describing • Drawing • Area, perimeter, and volume
Base 10 blocks	• Number sense • Place value • Addition, subtraction, division, and multiplication • Area and volume
Calculators	• Calculations • Algebraic equations • Problem solving • Mathematics facts • Statistics

Tools	Possible Uses in a Mathematics Classroom
Computers	• Number and operations concepts • Equations spreadsheets • Graphing • Problem solving • Algebra equations • Mathematics games
Dice	• Application games • Math-facts games • Addition, subtraction, division, and multiplication • Number sense • Prediction and statistics
Flash cards	• Numbers and operations concepts • Mathematics facts • Addition, subtraction, division, and multiplication
Fraction bars	• Comparing and ordering • Addition, subtraction, division, and multiplication • Number sense • Finding equivalent parts • Fractions
Geoboards	• Geometry • Sorting and classifying • Describing • Drawing • Symmetry • Spatial reasoning
Linking cubes	• Number sense • Addition, subtraction, division, and multiplication • Comparing and ordering • Area and volume
Number cards	• Numbers and operations concepts • Positive and negative numbers • Ordinal and cardinal numbers • Addition, subtraction, division, and multiplication

Tools	Possible Uses in a Mathematics Classroom
Number lines	• Number and operations concepts • Positive and negative numbers • Fractions • Ordinal and cardinal numbers • Addition and subtraction
Pattern blocks	• Number sense • Whole numbers • Algebra • Fractions • Proportions • Spatial visualization/estimation • Patterns • Representation of data and concrete objects
Play money	• Equations with money • Real-life situational word problems • Problem solving
Rulers (metric and standard)	• Measurement • Length • Standard and nonstandard units • Selecting appropriate tools of measurement • Comparing metric and standard measurements
Scales (metric and standard)	• Measurement • Standard and nonstandard units • Weight • Selecting appropriate tools of measurement • Comparing metric and standard measurements • Balancing equations in algebra

Effective instruction recognizes that students conceptualize mathematical and scientific concepts in different but often equally appropriate ways (Leinwand 2009). It is important to emphasize multiple representations and approaches to meet the learning needs of all students. The use of mathematical tools allows students to explore and concretely make sense of the mathematics.

Developing Mathematical Vocabulary

In mathematics, vocabulary is highly specialized—mathematics terms are not often encountered in everyday life. Therefore, all students need an explicit introduction to and explanation of these vocabulary words in order to apply them to their understanding of mathematical concepts. This task is even more difficult for English language learners, as these vocabulary words are not typically the words that English language learners will learn during their structured English language development (ELD) class period. It is up to the mathematics teacher then to ensure that English language learners study and learn the necessary vocabulary to comprehend mathematical concepts and curriculum.

Furthermore, the different areas of mathematics (e.g., number sense and mathematical reasoning) and the various disciplines (e.g., algebra, trigonometry, geometry, and calculus) have different compilations of specialized vocabulary words. Sometimes the words overlap across mathematical areas and disciplines, but often words are specific to just one mathematical area or discipline. It is vital to understand the vocabulary for a specific discipline in mathematics because this knowledge aids in access to the core curriculum. Students develop a deep understanding of mathematical language only through several approaches that develop mathematical concepts and connections (Dacy, Bramford-Lynch, and Salemi 2013). Teachers want students to demonstrate mastery of concepts, but even more important, teachers want students to think mathematically as lifelong learners. This will be possible if they first achieve understanding of the vocabulary words that explain, describe, justify, and facilitate each of the mathematical concepts.

Mathematics uses technical language. It is helpful if teachers use the technical language as often as possible for consistency and continuity among mathematics classrooms. The following sections give examples of how vocabulary can be developed in a mathematics classroom.

Vocabulary Development for English Language Learners

Mathematics teachers need to explicitly and deliberately teach the academic language needed for students to be successful in doing mathematics, without interrupting students' reasoning (Kanold et al. 2013). It is important for all students learning mathematics to be familiar with the specialized vocabulary embedded within the practice and application of the concepts. English language learners especially need consistent, structured instruction in learning English. They also need well-planned, sheltered instruction throughout content lessons and effective activities to develop mathematical vocabulary (Dean and Florian 2001).

It is not enough to give English language learners a list of words and have them look up the definitions in dictionaries or textbook glossaries. Students who are struggling with learning a language are not going to find the process easier by simply being given more words to sort through. English language learners need context-embedded lesson activities that acquaint them with the necessary words for comprehension of the content and allow them to practice the use of the words in activities that span listening, speaking, reading, and writing actions. The following example could help acquaint English language learners with the use of the term *symmetry* in relation to geometry in a support class or as a scaffold during the lesson.

Draw the lines of *symmetry* within the given shape:

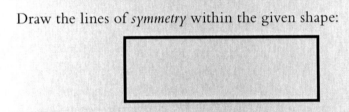

In a support class, the teacher could take time to review rectangle, line symmetry, and reflection terms. Then, the teacher could have students find the lines of symmetry of different shapes. Students may use a mirror, trace paper, or a MIRA (a tool that has the reflective quality of a mirror) to find the lines. During the support class, the teacher can encourage students to use—and even overuse—correct terminology as they work. In the support class or as scaffolding in the regular mathematics classroom, the teacher can have students reflect, orally or in writing, about the vocabulary and work that they accomplished, and help students see that the "reflections" mirror back their thinking. Connections like these are needed to help English language learners gain specific language.

Mathematics teachers need to be cognizant of the language difficulties students who are learning English may have. Many mathematics teachers believe English instruction is the job of the English teacher; however, the English teacher is not focusing on the specialized mathematical vocabulary and the contexts appropriate to it during English class. It is necessary for the mathematics teacher to offer the scaffolding students need to access mathematical concepts. By knowing the language level of each individual student, as described in the next sections, the teacher can plan appropriate lessons that balance vocabulary development, instruction, modeling, interactive activities, and support.

Types of Language Proficiency

One major concept mathematics teachers need to recognize is the difference between the two types of language proficiency for English language learners. Jim Cummins calls the two types of language Basic Interpersonal Communication Skills (commonly referred to as BICS) and Cognitive Academic Language Proficiency (commonly referred to as CALP) (Crawford 2004). BICS refers to a student's *social language*. Proficiency in social language requires no specific instruction and typically takes as little as three years to acquire. This knowledge can be acquired through media saturation, music, and social situations. Students can easily seem very capable in social language because they need it to survive. For example, a teacher may hear a student chatting with friends and converse with that student before or after class, which may lead the teacher to believe that the student has a firm grasp of the English language. However, that same student might be failing assessments, struggling to keep up with assignments, and unable to write well about mathematical content, indicating that the student lacks CALP.

CALP, or academic language, takes seven or more years to acquire. CALP is proficiency in the language of the content areas and of the classroom. A student who has strong CALP has a command of the use of English within content areas. In mathematics, a student with a strong level of CALP is able to understand key vocabulary, use it in the correct context, and write well about his or her understanding of mathematical concepts and procedures. This level of academic language is not learned easily and intuitively, like BICS. This language proficiency comes only with explicit instruction and planned objectives by the content teachers. That is one reason why vocabulary development lessons are so important for teachers of English language learners to incorporate into mathematics lessons.

Levels of Language Acquisition

Effective mathematics teachers of English language learners also need to know the levels of language acquisition for each English language learner in the classroom. The appropriate lesson for a student new to English is going to look very different from the appropriate lesson for an English language learner who is close to being considered fluent in the English language.

Many states have official assessments meant to determine the level at which a student is able to use English. These assessments cover the areas of listening, speaking, reading, and writing. Some of the assessments have a separate score for each domain of language and a composite score that combines the overall level at which the student is performing in English.

Figure 5.2 outlines the levels of language acquisition of English language learners and offers suggestions to teachers for how to meet those students' needs.

Figure 5.2 Levels of Language Acquisition and Teacher Use

Level	Description	Teachers Should...
Beginning	These students fall into a wide range of limited English comprehension. They have minimal or limited comprehension with no verbal production. Some beginning students are able to give just one- or two-word responses. Some are beginning to comprehend highly contextualized information and are able to speak in very simple sentences.	Provide a lot of context for mathematical concepts. Use physical movement and visuals to explain mathematical vocabulary. Use sentence frames to help students place mathematical concepts into context. Ask yes/no questions or questions in which the answers are embedded. Always include vocabulary development activities.

Level	Description	Teachers Should...
Early intermediate/ intermediate	These students have good comprehension of information in context. They may exhibit restricted ability to communicate ideas, but they can usually reproduce familiar phrases in simple sentences. As they improve in proficiency, their ability to communicate ideas improves, although they may exhibit errors in production, especially when writing or speaking about highly specialized content.	Provide visuals and context for mathematical concepts. Encourage cooperative and interactive activities in order to make mathematical content comprehensible. Ask questions that require simple sentences with known vocabulary. Elicit simple explanations and summaries. Support writing and reading tasks. Often include vocabulary development activities and the proper ways to communicate using the mathematical vocabulary.
Early advanced/ advanced	These students may appear to be fluent in English, but they often struggle when they have to explain their understanding of an answer or write out the procedures of a concept. They lack the ability to fully communicate higher levels of thinking in content-specific academic language.	Provide structured group discussion of concepts before requiring individual practice and writing about mathematical reasoning. Elicit explanations that analyze and synthesize mathematical information. Model the higher levels of thinking with use of specialized vocabulary. Regularly practice vocabulary development activities, and then take students to the next levels of higher-level thinking using the vocabulary.

Integrating Vocabulary Development into Instruction

Interactive vocabulary development activities should be regularly integrated in mathematics lessons in all classrooms. These types of activities are especially necessary for classrooms with English language learners, students struggling with mathematical concepts, or any students who have not shown mastery of the vocabulary.

Teachers should follow these guidelines before beginning to teach the vocabulary activities described in the sections that follow:

- Decide how long to use one vocabulary activity before introducing a new one.

- Plan for extra teaching time when a new vocabulary activity is being introduced.

- Choose an appropriate activity in order to meet the allotted classroom time for the particular lesson.

- Front-load the lesson with vocabulary words before students need to apply them during practice activities and problems.

- Revisit past vocabulary words in addition to current words, if a lesson requires them.

- Repeat the activity with the same words or new words if it needs to be practiced a few times before students can correctly perform the activity.

- Clearly state the purpose for an activity and student behavior expectations.

The activities described next often cover all four domains of language: listening, speaking, reading, and writing. It is important for students to practice using new vocabulary in a multitude of ways in order for them to take ownership of words and use them independently. Repetition is as important as exposing students to words in multiple ways.

During these activities, it is imperative for the mathematics teacher to actively monitor the process. The role of the teacher is to clear up any vocabulary misconceptions or misunderstandings. The activities can be used as front-loading activities, as practice centers for students to use with partners or small groups, or as remedial or enrichment activities during extra instructional time.

Activity 1: Chart and Match

The Chart and Match activity can be used for a lesson or an entire unit. From the pacing guide, the teacher finds the essential vocabulary for the unit, reviews the word with students, and gives an example or picture of the word. It is important for students to collaborate and construct their own definition of the word.

1. The teacher writes the vocabulary words specific to the day's or unit's lesson on the board.

2. Students create a three-column grid (see Figure 5.3) labeled with the following headings:

   - Word
   - Illustration or Example
   - Definition or Description

   Students write the vocabulary words down the left side of the grid. One term goes in each row.

3. The teacher introduces the words and leads a whole-class discussion. The teacher uses examples and draws pictures to show the vocabulary words.

4. The teacher directs students to draw a picture or write an example of the vocabulary word in the middle column of the grid, next to the corresponding vocabulary word.

5. In the final column, the class decides on a way to describe or define the word. This should not be a dictionary definition; rather, after the discussion, students can write their understandings of the word or the teacher can help them write student-friendly explanations.

6. After reviewing the finished grid, students cut up the squares. With partners, they can work together to place the three sections for each vocabulary word (the word, the illustration or example, and the definition or description) together.

7. The teacher walks around to monitor progress and clear up any misconceptions. With extra time, the teacher can use one grid for a whole-class activity.

8. The teacher hands each student one piece from one cut-up completed grid. If there are not enough pieces for everyone, the teacher can substitute previous vocabulary words or repeat vocabulary words.

9. Students walk around, read their pieces to other students, and trade cards.

10. The teacher directs students to stop, read their final cards, and find the other two students who have their matching components. Each group of three (one student with the word card, one student with the illustration or example card, and one student with the description or definition card) stand together and present the vocabulary word to the class.

Figure 5.3 Chart and Match Activity

Word	Illustration or Example	Definition or Description
Independent variable	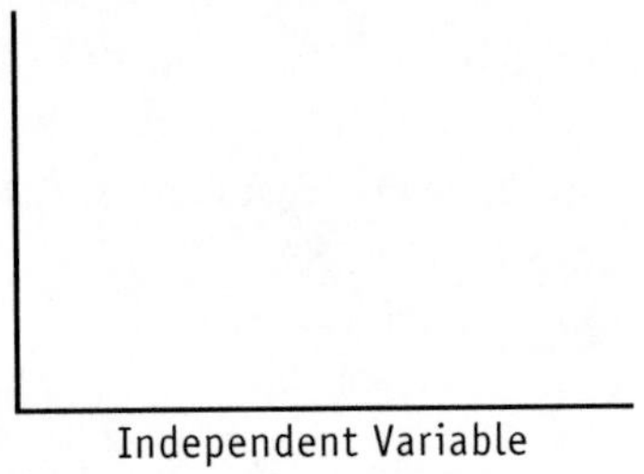 Independent Variable	The control or what I manipulate.
Dependent variable	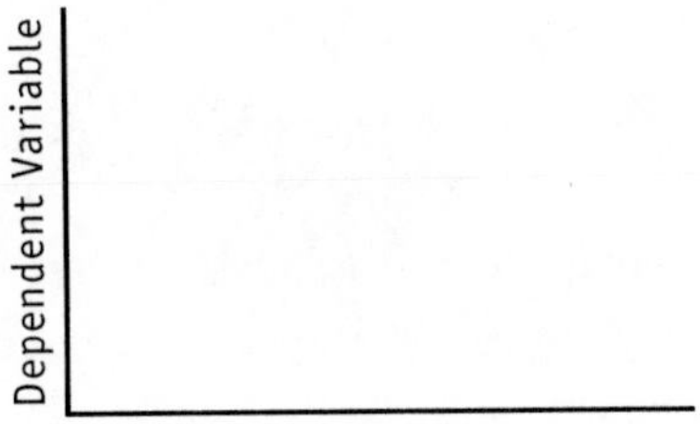 Dependent Variable	The result from what I manipulate.
Scaling the axes	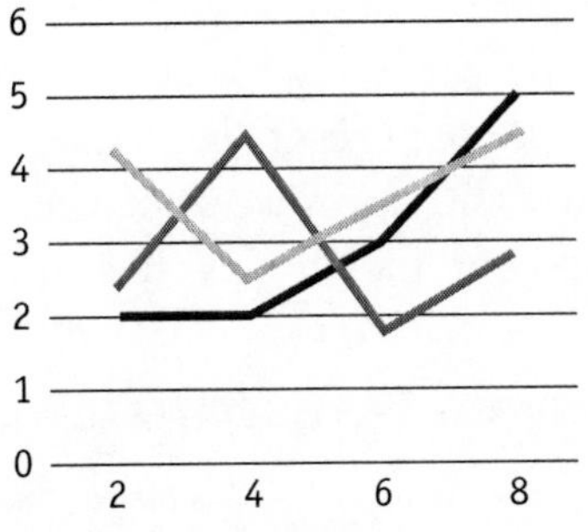	Numbers I count by, along the *x* and *y* axis. Must be equal intervals.

Word	Illustration or Example	Definition or Description
Histogram		Groups of numbers along the x axis. For example, height ranges.

Activity 2: Vocabulary Bingo

In this activity, students create their own bingo boards, which minimizes preparation for the teacher. This activity can be done individually or in pairs. The teacher follows the general guidelines/rules for bingo, adding mathematical vocabulary practice:

1. The teacher writes the vocabulary words specific to the day's lesson on the board (*polygon, quadrilateral, pentagon, hexagon,* etc.) and gives each student a blank grid.

2. Students are directed to write one word in each square.

3. The teacher starts the bingo learning activity by reading a description of the vocabulary word or showing an example or representative picture.

4. If there is extra time, the bingo winner can review the words and definitions out loud for the benefit of the whole class. Also, the teacher can have students review the definitions of their covered words with partners. This is a good time for whole-class summary.

Activity 3: Which Statement Is Inaccurate?

In this activity, each vocabulary word specific to the day's lesson is used in four written sentences. The teacher can do this ahead of time or ask students to write the sentences once they have been introduced to the vocabulary.

For each vocabulary word, three accurate mathematical sentences will be written and one inaccurate mathematical sentence will be displayed. (See the example that follows.)

1. The teacher displays four sentences. Three are accurate and one is inaccurate. The sentences are numbered one to four.

2. Students work in teams of four. Each team reads the sentences aloud.

3. During private think time, each student decides which sentence is inaccurate and makes note of his or her choice without letting the team see it.

4. When the team is ready, each student shows his or her choice to the team. The team discusses the answer in order to reach a consensus, writes the team's answer on a whiteboard, and is prepared to defend its choice.

5. When all teams have reached a consensus, the teacher asks the teams to display their answers to the rest of the class. The class can discuss each team's results.

6. If there is extra time, the teacher can have the teams convert the inaccurate sentence to an accurate sentence by making the appropriate changes.

Example: Which of the following descriptions of rectangles and squares is inaccurate? Be prepared to defend your choice.

1. A rectangle and square both have four sides.

2. A rectangle and square both have 90-degree angles.

3. A rectangle is a square and a square is a rectangle.

4. A rectangle and square are both quadrilaterals.

Solution: Statement 3 is inaccurate.

Activity 4: Frayer Model

Students work with partners using vocabulary terms or concepts and fill out the Frayer Model (see Figure 5.4). Students can create this model on a piece of paper or poster board, and the teacher should hang the models around the room so they are visible for the rest of the unit of study. Here are the instructions for the activity:

1. Students write a definition in their own words.

2. Students draw a picture or symbol that represents the word or mathematical concept.

3. Students create an example and nonexample of the word and put them in the appropriate boxes.

4. Students share their model with other groups within the class. The groups rotate until all groups have seen all models.

Figure 5.4 Frayer Model

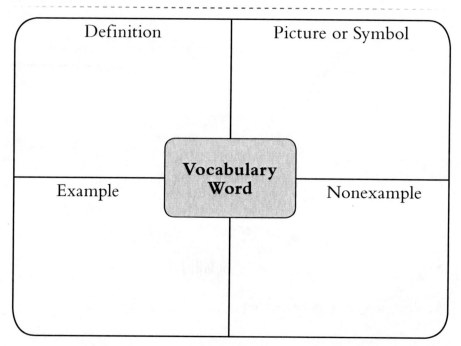

(Adapted from Frayer, Frederick, and Klausmeier 1969)

Figure 5.5 shows how the Frayer Model could be used in a mathematics classroom. The teacher should allow time for students to find definitions, examples, and nonexamples for the essential vocabulary.

Figure 5.5 Frayer Model in the Mathematics Classroom

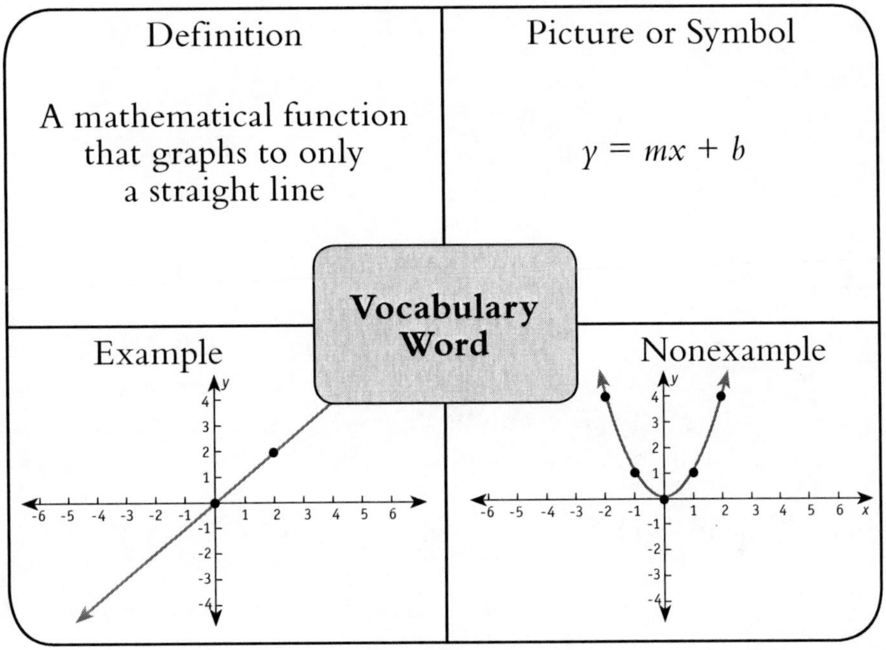

(Adapted from Frayer, Frederick, and Klausmeier 1969)

Activity 5: Sentence Frames for Vocabulary

This activity will vary depending on the vocabulary being introduced. It is best explained using a specific example, but teachers can adapt it to any set of mathematical vocabulary.

1. The teacher shares the vocabulary words specific to the day's lesson.

2. The teacher shares a simple sentence that frames the vocabulary in a proper mathematical context.

3. The sentence frame has blanks in which students will substitute information:

- The teacher should write the sentence frame on the board or on a sentence strip.

- The teacher models complete sentences using the vocabulary.

This example uses the sample vocabulary words *equation* and *equivalent*. When a teacher is teaching these words, a common way to express the concept is as follows:

The equation _____ is equivalent to the equation _____.

4. The teacher then writes sample answers to put in the blanks. For this example, the teacher writes:

$$3 + 5 = 8 \quad 6 - 2 = 4 \quad 4 + 0 = 4 \quad 9 - 1 = 8$$

5. Students work in pairs to orally rehearse the vocabulary with the right substitutions. Younger students benefit from having actual slips of paper to manually place into the blanks on their sentence strips. Then they can orally rehearse the information.

In the example, students would practice saying:

The equation $3 + 5 = 8$ is equivalent to the equation $9 - 1 = 8$.

The equation $6 - 2 = 4$ is equivalent to the equation $4 + 0 = 4$.

6. The teacher directs pairs of students to share the answers they came up with.

7. Partners determine other substitutions to correctly use with the sentence frame. They practice together and then share with the class.

Conclusion

It is important for students to investigate multiple approaches when solving mathematics problems. This chapter provided problem-solving processes that can be used in any mathematics classroom. It may be easier for teachers to give rote procedures for students to use when solving problems, but are students really learning? When students make sense of the mathematics and are provided time to think deeply, authentic learning happens.

Mathematics has technical language that cannot be ignored. Teachers need to take the time to teach the official mathematical vocabulary students need to know and understand. If students write the word and its definition, it does not follow automatically that they will remember what it means tomorrow. Students need to process the vocabulary in a methodical manner and practice mathematics talk with their peers. That is when the learning sticks.

Reflection

1. In what ways is it helpful to teach the problem-solving strategies? How will doing so help students become problem solvers?

2. How can mathematical tools help students explore math? Choose a lesson you are going to teach or have taught and find all the different tools that could help students construct meaning.

3. Why is developing mathematical vocabulary important? Identify two ways you can integrate vocabulary development into your instruction.

4. How can you discover the mathematical language ability of your students?

Chapter 6

Assessing Students

"In short, the effect of assessment for learning, as it plays out in the classroom, is that students keep learning and remain confident that they can continue to learn at productive levels if they keep trying to learn." (Stiggins 2002)

As teachers work through aligning assessments and standards, they should begin to ask themselves what student mastery should look like. This work should drive instruction. Summative and formative assessments provide valuable information to educators and families about a student's mathematical understanding. This chapter provides strategies, charts, and rubrics for summative and formative assessments, as well as ways to use data to further drive instruction in the classroom.

Aligning Assessment with Standards

To write effective assessments, teachers need to follow a process to align assessment with the standards. It is helpful if teachers work collaboratively when doing so, and professional learning communities (PLCs) offer this type of collaboration among teachers. During the process, it is helpful to unpack the standards, write learning targets and criteria for success, determine proficiency levels, write summative and formative assessments, and finally plan for instruction and differentiation. This ongoing process, shown in Figure 6.1, takes time and commitment to master.

Figure 6.1 PLC Planning Process

1	Unpack the standards
2	Write learning targets and criteria for success
3	Determine proficiency levels
4	Write summative and formative assessments
5	Plan for instruction and differentiation
6	Collect, analyze, and use data
7	Differentiate as needed

Each step in the process outlined in Figure 6.1 is important when designing assessments. The ultimate goal when writing assessments is to find out what students know and then how to fill in gaps, if any exist. Teachers should take time to unpack the standards. This information will be useful when writing assessments based on multiple levels of understanding. Teachers will use a collection of formative assessments, common formative assessments, and summative assessments in their classroom, as shown in Figure 6.2.

Figure 6.2 Assessment Types

Assessment Type	Description
Formative assessment	Individual teachers informally assess students' knowledge based on learning targets. These are assessments **for** learning that immediately inform instruction. • Exit or entrance tickets: Categorize triangles. • Thumbs up, down, or sideways: Can students explain least common multiple? • Observational data: Are students figuring out greatest common factors correctly? • Questioning students: Describe the characteristics of a regular polygon. • Activities focusing on mathematical vocabulary: Frayer Model. • Reflection on learning: Explain why a rectangle cannot be classified as a square.
Common formative assessment	A team of teachers informally assesses the data concerning students' knowledge based on learning targets. These are assessments **for** learning including all students within the grade level to inform instruction. • Give students addition problems with two-digit numbers to solve. • Have students solve problems using integers. • Ask students to build groups of tens using building blocks.
Summative assessment	Teachers formally assess students' knowledge based on the standards. These are assessments **of** learning at the end of instruction. • Solve real-world and mathematical problems involving perimeters of polygons, including finding the perimeter given the side lengths, finding an unknown side length, and exhibiting rectangles with the same perimeter and different areas or with the same area and different perimeters (CCSSM, 3.MD.D8).

There are distinct differences between formative assessments, common formative assessments, and summative assessments. Individual teachers use formative assessments to make instructional decisions each and every day. This can be done by a quick thumbs up, sideways, or down to determine student understanding of specific skills, procedures, or concepts. Grade-level teams write common formative assessments. Each teacher gives the common formative assessment to his or her students. The team reconvenes, analyzes the data, and plans instructional moves.

These types of assessments promote efficacy for teachers and students; help evaluate effective teaching strategies; inform the practice of teachers; build a team's capacity to improve its program; facilitate a systematic, collective response to students who are experiencing difficulty; and offer a powerful tool for changing adult behavior and practice (DuFour et al. 2010). Summative assessments are final measures of assessments against specific standards.

Mathematics teachers should consider the following when creating new assessments or reviewing previously created assessments to use with their students:

- The mathematics should be relevant and engaging to students.

- The expectations for the finished product should be clear.

- The assessment should clearly show students' mathematical knowledge, understanding, and thinking processes.

- The activity used for assessment should have a clear purpose for either formal or informal assessment.

When assessment aligns with instruction, both teachers and students benefit. Students are more likely to gain a deep understanding of the mathematics when instruction is focused and they are assessed on what they are taught. Instruction-aligned assessments are also time effective for teachers because they monitor learning and can be integrated into daily instruction and classroom activities.

Formative Assessment

Formative assessment on some level happens in mathematics classrooms throughout the country whether a teacher is aware of it or not. Instead of plowing through a pacing guide or scope and sequence, teachers intentionally change their instruction based on what they learn from students. The more intentional this becomes, the more vital a role formative assessment plays.

Paul Black and Dylan Wiliam (2010), authors of the article "Inside the Black Box: Raising Standards Through Classroom Assessment," found that the use of formative assessment can raise standards of achievement and help students make significant learning gains. Teachers should use the information gathered from formative assessments to adjust their pace or sequencing, determine content readiness, and provide feedback to students. If teachers give a formative assessment and do nothing with the data, then it is not a formative assessment—formative assessments always inform instruction.

To measure understanding, teachers assess students appropriately throughout each unit. Each day they provide students with a learning target. Sharing these learning targets with students is the foundational formative assessment strategy (Brookhart 2013). As students know what is expected of them, they will be able to judge whether they achieved the target. If the learning target states, "I can graph points in the first quadrant," the teacher is looking for students who understand the x and y values of an ordered pair. The teacher can also look for students who may not understand where the first quadrant is located.

Formative assessments should give teachers the information they need to make informed decisions about what to teach and how to teach it. The assessments must align closely with the mathematical concept being taught; it is important for teachers to appropriately use the information gathered from assessments. If not, teachers will not know whether students truly understand the concept. Asking students to graph ordered pairs will inform teachers of those students who understand, those who are close to understanding, and those who do not understand.

Formative assessment is a valuable way for teachers to get a quick understanding of how students are progressing in understanding a mathematical concept. However, given that these assessments are informal, teachers often do not record a grade; rather, they use the information gathered as data to guide instruction. *Formative assessment* is a relatively new term, but teachers have always been using information gathered from students to inform instructional decisions. Formative assessments are to be used as benchmarks along the way to the summative assessment. It is up to teachers to provide feedback and scaffold for students to understand a skill, strategy, or process.

For example, if a child begins swim lessons, he or she is not expected to immediately jump in the water and swim from one end of the pool to the other. There is a progression of learning that must happen before the child can swim the length of the pool. It is up to the swim teacher to set clear goals and provide feedback: put one's face in the water and blow bubbles, do assisted front and back float, do unassisted front and back float, and so on. The swim teacher will then measure the student's skills based on the goals and provide immediate feedback to the child. This feedback might include the specific skills to develop simultaneous and alternating arm and leg actions on the front and back. The swim teacher will then work with the child to master these skills. This example relates directly to formative assessment in mathematics education. A teacher sets a goal and helps students achieve the goal using all means necessary. For example, students cannot be expected to represent proportional relationships using equations if they do not understand fractions, unit rates, proportions, and writing equations. Learning is a progression that needs to be broken down into clear targets. Once these targets are identified, teachers can provide specific feedback to help ensure student understanding.

Teachers can make formative assessments at several points during a lesson:

- After accessing prior knowledge
- While students are exploring a task
- After the summary of a lesson
- During independent practice

Depending on the results, teachers can decide if they should continue with the lesson as planned or switch gears if student understanding is low. The next section presents ideas for checking for understanding.

Check for Understanding

When teachers check often for student understanding, they know whether to proceed with further lesson concepts, repeat instruction for some lesson concepts, offer more practice with the concept, or skip a portion of the lesson that students already understand. If a teacher notices a small group of students who do not understand the difference between mean, median, and mode, he or she can pull the small group to a table to clear up any confusion. However, if the entire class does not understand the difference between mean, median, and mode, the teacher will need to "catch" the entire class and have a class discussion about the measures of central tendency to sort out any misunderstandings. It is okay to use this knowledge to differentiate instruction by allowing some students to work independently while the teacher works with a small group of students who need more instruction. Overall, checking for understanding shows how to adjust instruction so that the lessons directly meet students' instructional needs (Wiliam 2007).

Figure 6.3 shows ways teachers can check for student understanding. These are quick formative assessments that teachers can use to help inform instruction.

Figure 6.3 Strategies for Checking for Understanding

Goal	Strategy	Summary of Strategy
Pacing the lesson Mathematical understanding	Display thumb signal	The teacher asks students to show a thumb signal. Students put their thumbs up or down, or waver them in the middle to demonstrate the following: "Yes, I totally understand this," "No, I do not understand this," or "I think I understand this."
	Fist to five	Students may indicate their understanding by showing a fist (I have little to no understanding) or up to five fingers (I totally get this and can teach it to others).
	I used to think...Now I think...	Students finish the phrase either in writing or orally with a partner.
	Exit tickets	The teacher provides a question to the class and students individually write their response on a sticky note or notecard. When they turn in the response, they may exit class.
	Murky math	At the end of class, students write down what they thought was most difficult or confusing.
Engaging and motivating students Activating student discourse	Pair-share	Students solve a problem and share it with a partner. When both partners agree on the solution, they show the teacher a thumbs up signal.
	Response cards	The teacher types questions onto cards relating to a mathematical concept and puts them in an envelope. Students draw a card and work with a partner to read and answer the questions.

Goal	Strategy	Summary of Strategy
Formalizing mathematical concepts	Whiteboards	Students complete a problem on a whiteboard and show their answer to the teacher. The teacher can quickly gauge whether students are solving the problems appropriately and can then make lesson decisions accordingly.
	Example/ nonexample	Students write an example and a nonexample of the mathematical concept.
	Fast writing assignment	Students quickly write two or three sentences explaining the concept or sequencing the procedure for solving a problem.

Again, formative assessments require teachers to immediately do something different with their instruction. Generic instructional strategies may not be as helpful as mathematics-specific strategies (Keeley and Tobey 2011). The teacher must uncover confusion and work to improve mathematical understanding. Formative assessment is a time where students can show what they know, and it is also a time for teachers to provide feedback on specific skills or strategies to help students improve as mathematicians. Individually, teachers check for understanding, but an even more powerful approach is the common formative assessment. This involves teachers being willing to take risks with their colleagues. This is the value of a PLC.

Teachers in a PLC create a common formative assessment based on targets or small benchmarks. Even if the first attempt at creating a common formative assessment does not give the teachers the data they were hoping for, they should try again. It takes time and usually multiple attempts at writing a quality assessment to receive informative data. When writing a high-quality common formative assessment, teachers should consider the following questions:

- Does the common formative assessment assess what we want to assess?
- What depth of knowledge are the questions?
- Is there a progression from easy to difficult?
- Is the assessment too long? Too short?
- Is there a balance between selected response and constructed response?

- Are the questions and directions clear?

- Is the font easy to read?

- Does the rubric align with the assessment? Does each learning target have its own criteria?

Teachers should create and give common formative assessments during every mathematics unit and use a rubric to evaluate student answers. All teachers should bring student work to their PLC meeting to examine the responses. It is helpful to work in teams to identify areas of success and areas that need improvement. When working with other teachers and examining student work, collaborative conversations start to happen and the focus tends to shift from the teacher to student learning. It may be intimidating and scary at first to share this information, but open and honest communication between teachers is vital to helping students succeed.

It is important to have easy methods of recording formative and common formative assessments. This information will help teachers to group students based on need or enrichment. It will also help teachers observe growth over time or continuous gaps in learning, so they are able to make informed teaching decisions.

Recording Methods

Teachers can use easy methods for recording informal assessments that do not take much additional class time to complete, such as the following. Clipboards or small binders allow for easy portability of the recording sheets.

- The teacher uses a blank grid to record notes for each student during a discussion, a pair activity, or small-group work.

- The teacher designs a check-off sheet with students' names down the side column and a list of several learning targets or concepts being taught in the lesson across the top. The teacher gives students a check, and writes the date and what was observed in each area when students demonstrate mastery throughout the lesson.

- The teacher posts a schedule of students he or she will meet with individually each day of the week. The teacher has an activity or quick task students can complete while at the meeting. He or she uses a data sheet to record any misconceptions and observations while each student is working.

- The teacher uses a general rubric to gauge student understanding. He or she walks around the room to observe and record understanding for several students each day to see how they are progressing.

As the goal is success on the summative assessments, this recording of student understanding will accurately show students' progress with mathematical concepts.

Summative Assessment

Two summative assessments usually given by school districts are benchmark assessments and state assessments. *Benchmark assessments* are district–level assessments given to students after each unit, once each quarter or semester, or yearly. These assessments are usually given to monitor and predict student achievement on the state assessment and to create a balanced and coherent assessment system throughout the district. Many states have adopted Partnership for Assessment of Readiness for College and Careers (PARCC), Smarter Balanced Assessment Consortium (SBAC), or they are crafting their own state assessment. These consortiums have created a common set of assessments in English and mathematics aligned to the Common Core State Standards (CCSS). State assessments are usually used to determine proficiency percentages in each school, identify areas in which groups of students may be underperforming, or make programmatic and placement decisions. For example, teachers and administrators analyze summative data to make decisions about placement in mathematics review or accelerated mathematics classes.

Written tests and quizzes are often the most common type of summative assessments used in classrooms. These types of assessments take place after all instruction has ended. They report the final results of student learning to students, teachers, parents, and administrators. Reporting of summative assessments occurs with either a letter grade or level of proficiency based on mathematics standards—for example, how eighth-grade students understand the Pythagorean theorem. If the summative assessment is the last time students will be exposed to a certain standard, the results will not be used to improve student understanding. The formative assessments teachers used throughout the unit of instruction typically provide the feedback necessary to be successful on the final summative assessment. Rubrics, explained in the next section, are effective clarifying instruments that can be used in both formative and summative assessments.

Rubrics

According to Myron Dueck (2014, 84), "Rubrics are fantastic formative assessment tools, as they provide students with the opportunity to identify areas needing improvement before the final grading stage." Rubrics are a way to provide feedback to students about skills or performance. Heidi Goodrich Andrade (2000, 13) states, "Rubrics make assessing student work quick and efficient, and they help teachers justify to parents and others the grades that they assign to students." This section explains the effectiveness of rubrics and gives examples of different rubrics that mathematics teachers can use in the classroom.

Rubric Effectiveness

The use of rubrics can be an effective assessment strategy in a mathematics classroom. Teachers may want to create general rubrics to aid in the process of using assessments to direct further instruction. With rubrics, teachers can create and convey realistic expectations for student work by showing exactly what learning students should demonstrate. Rubrics are helpful for teachers as well. Focusing on student learning through rubrics will help educators think about the outcomes rather than tasks.

When a teacher communicates clear learning targets and then provides a rubric that delineates the expectations of the completed assignment, students are supplied with the necessary information to be successful with an assignment. Without clear learning targets, there is nothing to assess. Without clear learning targets, students do not know what the purpose is (Brookhart 2013). Rubrics can be especially useful for assignments where students have to explain their reasoning or their comprehension of a particular lesson concept. In this case, the teacher could create one general rubric that could be used many times throughout the year.

A well-written mathematics rubric contains the overarching learning goal and performance-level descriptors. Teachers need to teach students how to read and analyze rubrics before beginning assignments, check their work throughout the assignments, and finally use the rubric guidelines to evaluate their evidence of learning. It is important for teachers not to use rubrics as a checklist of what students need to do (write his or her name on the paper, color in the diagram, answer five of the seven questions, etc.), but as criteria of *what* students will learn.

Creating a Mathematics Rubric

Before creating a mathematics rubric, teachers need to choose the criteria they are measuring. State and district standards are great places to begin when creating a rubric because they reveal exactly what needs to be mastered. Then teachers can follow these steps:

1. Identify performance-level descriptors for the criteria.

2. Generate a list of the concepts students are expected to master (content standards) as well as another list of ways they can show that they comprehend the material (process standards).

3. Highlight the most important items on both lists.

4. Use the content standards and process standards to describe the performance criteria in detail. These should identify the learning outcomes.

It is important for teachers to decide how many criteria to include. The more detailed the rubric, the more involved the scoring process. Many teachers find that three to five items are ideal. Too few and the teacher will be unable to effectively evaluate the performance of students with varied learning abilities. Too many and the rubric will be too extensive and time consuming to use. The rubric should be easily comprehensible for students so they can evaluate their own performance.

Typical headings can be a 4, 3, 2, 1 scale or the words Advanced, Proficient, Nearing Proficient, and Novice. It is important to have clear criteria for each performance level. Figure 6.4 shows generic language that can be used when creating a rubric for a learning objective. Teachers need to make sure to use specific language that matches standards when creating their own rubrics.

Figure 6.4 General Overview of Mathematics Rubric Guidelines

Advanced Level: 4 or Advanced	Highest Level: 3 or Proficient	Middle Level: 2 or Nearing Proficient	Lowest Level: 1 or Novice
Understands skills, procedures, or concepts and extends this understanding beyond the requirements Uses mathematical terminology or notations accurately Process complete beyond expectations Clear explanation	Understands skills, procedures, or concepts Uses mathematical terminology or notations accurately Process complete Clear explanation	Basic understanding of skills, procedures, or concepts; lowest acceptable score Uses some mathematical terminology or notations Process complete but contains errors Unclear explanation or parts omitted	Minimal comprehension of skills, procedures, or concepts Uses little to no mathematical terminology or notations Process not complete Little to no explanation

Not all rubrics will have the Advanced column. The teacher should try to offer advanced thinking and learning to students, but sometimes Proficient or level 3 is all students will be able to achieve with certain learning goals.

Figure 6.5 outlines general criteria for various steps of an assignment, including completion of problems, correct calculations, answers that relate to the topic, and logical reasoning. This type of rubric can be useful when reviewing homework or in-class work.

Figure 6.5 Three-Point Mathematics Rubric

Points	Criteria
3	The solution is correct and the student has demonstrated a thorough understanding of the concepts, procedures, and skills. The task has been fully completed using sound mathematical methods. The response may contain minor flaws, but it is thorough and the understanding is evident.
2	The response is only partially correct. The solution may be correct, but the response demonstrates only a partial understanding of underlying mathematical concepts, procedures, and skills. Or the solution is wrong, but the student demonstrates understanding of the concepts and procedures.
1	The solution is incorrect. The response is incomprehensible and/or demonstrates no understanding of the concepts.

(Adapted from *FCAT 2004 Sample Test Materials,* Florida Department of Education 2003)

Figure 6.6 shows a sample rubric from the Grade 6 Ratios and Proportional Relationships Standard from the CCSS. (Please note that this is not the entire standard, but a part of the Ratios and Proportional Relationships Standard.)

Figure 6.6 Ratios and Proportions Rubric for Sixth Grade

	Ratios and Proportions			
	Advanced	**Proficient**	**Nearing Proficiency**	**Novice**
Proportional relationships	Solves real-world problems involving proportional relationships that require one step with measurement conversions	Represents proportional relationships in graphs and tables, and solves one-step rate-related problems	Identifies proportional relationships presented in equation formats and finds unit rates involving whole numbers	Finds missing values in tables that display a proportional relationship
Solution process	Shows all the steps when solving the problem	Shows some of the steps when solving the problem	Shows few to no steps when solving the problem	Does not show steps
Clarity of explanation	Completely explains mathematical work	Mostly explains mathematical work	Writes some explanation, or explanation does not make sense	Does not write any explanations

(NGA and CCSSO 2010)

Most mathematics rubrics should include mathematical procedural and conceptual knowledge, strategic knowledge, and mathematical communication (Brookhart 2013). These three areas combined will help enforce the standards and require students to know and communicate about mathematics.

Responsibility, effort, study skills, work habits, homework completion and quality, class participation, punctuality in turning in assignments, attendance, and other similar aspects are critical to learning, yet should be reported separately from academic achievement (Guskey and Bailey 2010). If teachers are using rubrics to measure success in a mathematics class, they should also use rubrics to help describe other aspects that are critical to learning. For example, the rubric shown in Figure 6.7 describes learning qualities that all students should demonstrate in order to be successful in all classrooms.

Figure 6.7 Learner Responsibilities

	4 Consistently Exceeds Expectations	3 Consistently Meets Expectations	2 Inconsistently Meets Expectations	1 Does Not Meet Expectations
Independent practice **Homework**	Consistently attempts the problems and provides evidence of mathematical thinking	Usually attempts the problems and provides evidence of mathematical thinking	Sometimes attempts the problems and provides some evidence of mathematical thinking	Rarely attempts the problems or provides evidence of mathematical thinking
Participates in learning **Collaboration**	Consistently shares information or ideas when participating in discussion or groups Regularly uses teamwork and leadership skills to help and encourage others	Usually shares information or ideas when participating in discussion or groups Usually uses teamwork and leadership skills to help and encourage others	Sometimes shares information or ideas when participating in discussion or groups Sometimes uses teamwork and leadership skills to help and encourage others	Rarely shares ideas May refuse to participate In groups, relies on the work of others Rarely uses teamwork and leadership skills to help and encourage others
Follows classroom expectations	Consistently follows classroom expectations and routines	Usually follows classroom expectations and routines	Sometimes follows classroom expectations and routines	Rarely follows classroom expectations and routines

Data-Driven Instruction

Data-driven instruction refers to the process of designing curriculum and instructional strategies to match data from student assessments. This data can be collected from various daily activities such as student-teacher interaction and observations, guided and independent practice, and formative and summative assessments. The key to data-driven instruction lies in gathering and interpreting the data in a mathematics classroom and understanding how to use it.

Data analysis is not useful if it does not result in meaningful instructional change. Data-driven educators are able to use formative and summative assessment data together to implement strategic, targeted, focused instructional interventions to improve student learning. These interventions should be aligned with state or district standards and curricula as well as content-specific, developmentally appropriate best practices (McLeod 2005).

Once the data are collected, the results need to be analyzed. Data can be analyzed by student or by skill, concept, or standard. When looking at data from a particular student, it is important to see what skills the student is not mastering and any commonalities those skills possess. For example, if the student is having a difficult time multiplying and dividing, he or she is probably weak in addition and subtraction because those are the foundational skills needed for multiplication and division. Identifying the mathematical concepts with which the student is struggling will guide future intervention for that particular child. When looking at data by concept, skill, or standard, a teacher can see where a group of students is having difficulty. Depending on the number of students having difficulty, the teacher can create an intervention with just that particular group of students or reteach the concept to the whole class (Dean and Florian 2001).

The sections that follow describe how to collect, analyze, and use data to make informed instructional, mathematical decisions. It is important for teachers to collect baseline data on students. From there, they can take time to analyze the data so they have a clear understanding of what it means. These choices will be used to make informed instructional changes to best meet student needs.

Collecting Data

Teachers have massive amounts of data they collect on a daily basis. At times, it is too overwhelming to know what to do with all of it. It is truly beneficial to work collaboratively in a PLC to sort through all the data. Teacher teams should start by collecting data from the state test and district assessments. Using this data, teams can then recognize patterns or areas in which students are not succeeding.

When organizing data, teachers should use a simple method. If teachers choose a complex method, they will usually quit the process altogether. Bruce Wellman and Laura Lipton (2004) developed a quick and easy protocol to organize data called Here's What! So What? Now What? (see Figure 6.8).

Figure 6.8 Organizing Data: Here's What! So What? Now What?

Learning Target: _____

Criteria for Success: _____

Here's What!	So What?	Now What?
• Specific facts and data • What stands out? • What do we see?	• Conclusion • Why do we think this happened? • Interpretations and perspectives	• Implications • What are we going to do?

Analyzing Data

When teams have identified one or two areas in which students' scores are consistently low, it is helpful to create SMART goals. Jan O'Neill and Anne Conzemius (2005) created SMART goals for educators to use to analyze data. These goals focus on a few mathematical standards in which students are not succeeding and help teachers stay focused by setting realistic and attainable targets for students to master. The acronym SMART stands for:

- **S**trategic and specific: A clear learning purpose for students.
- **M**easureable: Concrete criteria to measure progress.
- **A**ttainable: Realistic timeline to achieve the goal.
- **R**esults oriented: Focus on student results and next steps.
- **T**ime bound: When will the data be collected? How often?

After teachers analyze the data from state tests, district assessments and, in some cases, classroom assessments, they then determine what the current reality is. It is important for teachers to work collaboratively in this process; otherwise, teacher biases and subjectivity might cause false assumptions. After defining the current reality, teachers brainstorm possible causes of the current reality. Teachers then work collaboratively to develop and implement a SMART goal action plan, as shown in Figure 6.9.

Figure 6.9 SMART Goal Action Plan

School: _____
Team Members: _____
Date: _____

Current Reality	SMART Goal		
Possible Causes Between Gap and Reality	**Strategies and Action Steps**	**Target Date or Timeline**	**Evidence of Effectiveness**

(Adapted from DuFour et al. 2010; Bailey and Jakicic 2012)

Using Data

Data by itself is not inherently useful. It is not until teachers put the data into a workable document, such as the Here's What! So What? Now What? chart shown in Figure 6.8, and analyze the results that the data become something they can use in the classroom or for interventions. Teachers can use the data to communicate with students as well as to identify and address student misconceptions to guide future instruction. For example, when a teacher is assessing whether students understand the concept of a ratio and a unit relationship, he or she can create a short assessment. If students demonstrate a solid understanding, the teacher will provide enrichment activities. If students do not understand the standard, the teacher must find the misconceptions and work with students to solidify understanding.

Communicating with Students

It is important to show students assessment data in order to help them understand their progress throughout the year. When students have data that measure their performance on a small task, they can see specific ways to improve their mathematical understanding. This gives students workable, meaningful goals. When students gain mastery of a mathematical skill, they are more confident and motivated to achieve in other areas of mathematics (Wiliam 2007). One way for students to track their understanding is by developing a graph where the independent variable is the learning target or standard and the dependent variable is their level of understanding based on a four-point rubric, such as the example shown in Figure 6.10. Students record their level of understanding. The data tracking can be used by the teacher, student, and guardians as a communication tool so all parties have a clear understanding where the student is in terms of learning.

Figure 6.10 Level of Understanding Ratios

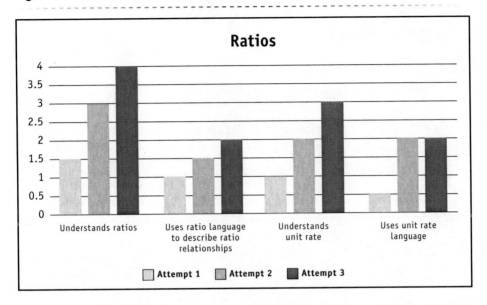

Students do not always understand why they are having difficulties in a mathematics class. They might realize that they are struggling but feel powerless to "catch up" once they fall behind. They might claim that they "do not like math," when in fact they do not understand many mathematical concepts. Sometimes when students feel discouraged, they react angrily in an effort to hide a sense of failure. While a teacher might find some students who actively seek a path for understanding, the teacher will also have students who do not know how to do so.

The following are some practical strategies for identifying student misconceptions and understanding.

Identifying Student Misconceptions and Understanding

1. The teacher uses frequent formative assessments to check for understanding during lessons.

2. The teacher carefully watches individual students for evidence of mastery or confusion.

3. When the results of the whole class are analyzed, the teacher uses diagnostic tools to help identify which concepts need the most instructional time.

4. The teacher keeps accurate records of student and class progress and uses them to see where students are struggling.

5. The teacher carefully monitors classroom and homework assignments for signs of how well students are grasping the content.

6. The teacher reviews student notes taken in class. The teacher can periodically ask students to explain their comprehension of vocabulary, procedures, or concepts in order to address student misconceptions in further lessons or decide if reteaching the lesson is appropriate.

7. The teacher uses rubrics. With a rubric, the teacher can identify the skills, procedures, or concepts evident or lacking in student work. The teacher can share rubrics with students or have students identify the rubric scores on their own papers.

Once the teacher has given the assessments, collected and analyzed the data, and identified any areas in which students are faltering, the teacher creates an action plan for addressing student misconceptions and reteaching materials. It is always more effective to change the strategies and instructional plans when reteaching a concept that students did not understand the first time. Teachers can also work collaboratively with a partner teacher and group all students based on need.

Reteaching Concepts

- The teacher can **choose a strategy from the differentiation chart** (in Chapter 7) to use for reteaching the concept. The teacher should choose the appropriate level from the chart to select an effective strategy for each group of students.

- The teacher can allow students **opportunities to talk about the concept** with partners or within small groups.

- The teacher can **choose an activity** such as **a game or a project** that further enhances students' understanding but presents the information in a new way.

- The teacher can allow those who successfully mastered the content to review different application activities while **meeting separately with a small group** of those who did not master the content.

- The teacher can **review the vocabulary with activities** if the teacher suspects that students still do not have the academic vocabulary necessary to comprehend the concepts and practice the skills.

Data can be used to convey information to parents, administrators, and the community as well.

Communicating with Parents

Data can be used as an effective tool in parent-teacher conferences and communications. The information allows parents to see specifically where their children are struggling or excelling, and it provides useful feedback for ways in which parents can help at home. The data will clarify what is expected of their children and where students are actually performing according to the state or district standards.

Communicating with Administrators

Data can be used to show administrators how students are progressing throughout the year in relation to mastery of state or district standards. Administrators will gain an understanding of how students may perform on any state-mandated, standardized assessments. The data will also indicate if or when during the year to begin an intervention program before or after school.

Communicating with the Community

In many states and districts, data are a measure of accountability for students, parents, teachers, and administrators. When the data reveal an overall trend in a program, school, or district, this is valuable information to show what is working and what is not working in the community's schools.

Student Self-Assessment

One factor often overlooked when using assessments is the student self-assessment component. Self-assessment does take some time to teach, as students may not have been asked to self-assess before. This reflection component will help students take an active role in their own quest for learning.

Students should be given multiple opportunities to demonstrate understanding of the concept as well as multiple opportunities to self-assess. Teachers write the learning targets so all students can see them, but teachers also need to give students time to understand what it is they are supposed to do. Teachers can have students talk with one another about the learning target and record the learning target in their own words. At the end of class, students can rate themselves in terms of understanding. Students begin to actively take part in their own learning when they self-assess.

Figure 6.11 shows a few examples of student self-assessments.

Figure 6.11 Student Self-Assessment Rubrics

Learning Target	I totally get this!	I almost get this!	I am SO lost!
I can read and write decimals to thousandths.			

Learning Target	😀	😕	😟
I can tell time.			

Learning Target	I can explain this to my friend.	I can almost explain this to my friend. I might need a little help.	Please, someone explain this to me!
I can graph points in all four quadrants on a coordinate plane.			

It takes time for students to master a learning target, and more often than not, it takes more than one day. Offering an example of the learning target is an alternative to assessing the learning target by itself. Teachers could give a rubric with all the learning targets and examples before a unit is even taught, and provide opportunities to students to read through the examples and assess themselves periodically throughout the unit. Figure 6.12 shows the start of a sample rubric for a seventh-grade unit involving the number system.

Figure 6.12 Learning Target Rubric for Seventh Grade

Learning Target	Example	I get it!	I almost get it!	I have lots of questions!
I can represent addition and subtraction on a number line.	$-5 + 2$			
I can solve real-world problems using addition, subtraction, multiplication, and division.	Sarah's cell phone bill is automatically deducting $64 from her bank account every month. How much will the deductions total for the year? *Solution:* $-64 + (-64) + (-64) + (-64) + (-64) + (-64) + (-64) + (-64) + (-64) + (-64) + (-64) + (-64) = 12 (-64)$			

Conclusion

Intentional teaching ensures student mathematical understanding. Teachers and students need to understand the purpose of a mathematics lesson and be able to assess whether they hit the target. Formative and common formative assessments are ongoing valuable tools for making sure that learning happens and student growth occurs. With effective checks for understanding and clarifying tools like rubrics, students can be guided to a higher level of achievement on a summative test.

Reflection

1. Explain the importance of this continual process:

    a. Write learning targets and criteria for success.

    b. Determine proficiency levels.

    c. Write formative and summative assessments.

    d. Plan for instruction and differentiation.

2. In what ways can formative assessments inform instruction?

3. Create a rubric that assesses a learning target based on a standard you will be teaching this year.

Supporting Instruction Through Differentiation

As teachers get to know their students, they realize that all students have unique personalities and behaviors. When teachers get to know their students as *mathematics* learners, they realize there is not a one-size-fits-all approach to teaching and learning. Students have diverse learning styles, come from different cultures, have varying levels of language ability, and differ in mathematical readiness. Students may bring a fear of mathematics to the classroom, as their parents may have had poor experiences in their own mathematics classrooms, and there is a pervasive cultural view of mathematics class as a form of punishment. Students may even have heard teachers of other subjects say, "I don't do mathematics."

It is easy to see how complex a mathematics classroom can be. Because of this complexity, teachers have realized that they must differentiate their teaching to better meet these diverse needs. Rick Wormeli (2006, 3), the author of *Fair Isn't Always Equal: Assessing and Grading in the Differentiated Classroom*, offers this definition of *differentiation*:

"Differentiated instruction is doing what's fair for students. It's a collection of best practices strategically employed to maximize students' learning at every turn, including giving them the tools to handle anything that is undifferentiated. It requires us to do different things for different students some, or a lot, of the time in order for them to learn when the general classroom approach does not meet students' needs. It is not individualized instruction, though that may happen from time to time as warranted. It's whatever works to advance the students. It's highly effective teaching."

Differentiation in a mathematics classroom has many faces depending on the particular teachers and students involved, the outcomes of these learners, and the structure of the classroom environment (Pettig 2000). Differentiation encompasses what is taught, how it is taught, and the products students create to show what they have learned. At the core of the classroom practice of differentiation is the modification of four curriculum-related elements—content, process, product, and affect—which are based on three categories of student need and variance—readiness, interests, and learning profile (Tomlinson and Imbeau 2010):

- **Content:** What teachers teach
- **Process:** How students learn
- **Product:** The final outcome
- **Affect:** The emotional impact

If a learning experience matches closely with a student's skills and prior knowledge (readiness), he or she will have more access to learn. Creating mathematics assignments that allow students to complete work according to their preferences (learning profile) will help learning experiences become more meaningful. For example, a teacher could take students on a school-wide scavenger hunt recording different shapes. If a topic sparks excitement in students (interest), then they will become more involved in learning and will better remember what they learned. Generating interest can be as simple as including student names within the word problems, or the teacher asking, "Could a dinosaur fit into our parking lot? Let's find out!"

To make mathematics activities most effective, teachers should take time to think about all the ways differentiation will help learners' progress. Not all students need to be engaged in exactly the same activity at exactly the same time. Differentiation is absolutely necessary to ensure the success of all students. It is the teacher's responsibility to continuously monitor student progress and to provide interventions as needed.

Through formative assessment, summative assessment, and diagnostic data, teachers can find out quickly where a student struggles and intervene immediately. It is helpful to collaborate within a professional learning community (PLC) to identify what students need to understand and do. The PLC team will plan for interventions before a unit is even taught. The task of collaborative team members is to become experts at implementing the strategies, routines, and interventions that will result in improved student learning (Kanold et al. 2013).

This chapter covers different strategies and techniques that provide all students with equitable opportunities to access mathematics. For all students to learn at high levels, it is helpful to provide multiple instructional intervention settings and strategies, as described throughout this chapter. Basic instruction to all students must be solid because no system of intervention will compensate for bad teaching (DuFour et al. 2010). Schools need to be prepared for students who have gaps in their learning and must have a plan to fill those gaps.

Response to Intervention

All students learn at different rates, so teachers need to have planned interventions to best meet students' needs. Many of the same researchers who created the Reading First initiative developed a system of identification known as Response to Intervention (RTI). The RTI model supports the idea that teachers should look for curricular intervention designed to bring a child back up to speed as soon as he or she begins having problems. According to Greg Cruey (2006), "RTI has the potential then to allow disabilities to be identified and defined based on the response a child has to the interventions that are tried." Depending on the level of difficulty they are having with the mathematics curriculum, students are classified as Tier 1, Tier 2, or Tier 3.

For students struggling in mathematics, it is vital to provide interventions as soon as possible so they do not fall behind. According to RTI, not all students need the same type of intervention. The teacher should assess students' needs and make proper recommendations immediately. For example, Tier 1 students need interventions as part of daily classroom mathematics activities. Teachers of these students should be able to complete this type of intervention within their own classrooms using differentiated instruction, strong mathematics tasks, multiple entry points, emphasis on vocabulary, guided instruction, scaffolding tasks, and concrete models as a reference. Tier 2 and Tier 3 students need more intense and focused help than a classroom teacher can provide on his or her own within the mathematics class. (Figure 7.2 shows this in more detail.)

Intervention is another area where a PLC can be extremely helpful. Teachers in a PLC should meet frequently to design intervention programs. Ideally, these meetings occur at the end of the school year to plan for the following year, during the summer, and at the very beginning of the current school year. The sooner decisions are made regarding the programs that will be implemented, the more time there is to develop the timeline, plan instruction to meet students' needs, and conduct professional development. Considerations include the following:

- Which type(s) of intervention will be offered

- The amount of instructional time for each type of intervention

- The length of each program

- The specific curriculum that has been chosen and the timeline for teachers to follow

- How to assign and train personnel

- The criteria to move in and out of an intervention program

- The fluidity of the program (i.e., Can students leave once they have achieved growth or must they stay until the end of the quarter? What are the criteria to move in and out of an intervention program?)

Key Points to Remember When Using an Intervention Program

When teachers differentiate instruction based on need, they should be flexible with their teaching and with their students. The following helpful reminders are for teachers working with students who may need more assistance and time:

- Use concrete objects to help students understand abstract representations and notations.

- Provide clear problem-solving models.

- Model thinking aloud and have students do the same.

- Include multiple representations of concepts.

- Allow students to move through the stages of mathematical development (concrete, abstract, application) slowly so connections are made between each step.

- Create an environment in which students feel comfortable asking questions and discussing concepts when they do not understand something.

- Keep open communication among teachers, administrators, students, and parents so everyone understands what progress is being made and what learning still needs to be developed.

- Help students choose learning goals to keep them motivated throughout the intervention program.

These reminders help teachers provide students with more assistance and time. Teachers should keep in mind several other considerations when differentiating instruction:

- Show individual student progress that is data-driven based on essential standards, mathematical disposition, and work habits.

- Allow an appropriate amount of time for planning and preparation of instruction and activities.

- Assess students often and use the results to guide instruction and appropriate intervention strategies, if necessary.

- Make specific plans to meet varied student needs and communicate those plans clearly to parents.

- Use many flexible grouping strategies throughout classroom activities.

- Vary the types of activities and modes of instruction used in the classroom to meet the learning needs of all students.

- Continually reflect on personal teaching strategies and modify the methods of mathematics instruction if students are not responding positively to the delivery methods.

- Ask for help from specialists, other mathematics teachers, parents, administrators, and anyone else who can assist in responding to student needs.

The following section identifies the breakdown of time in terms of acceleration, remediation, and recovery when planning for differentiation.

Implementing Strategies in an Intervention Program

When using an intervention, if teachers tend to focus only on skill development, they miss opportunities to close and fill in learning gaps. For example, when teachers plan their week using three days of acceleration, one and a half days of remediation, and a half day of recovery, the time allotment will help front-load grade-level mathematical concepts, fill in skill gaps, and recover any missing assignments. Using the three-pronged approach to mathematics intervention shown in Figure 7.1, teachers will clear up misconceptions and solidify learning.

Figure 7.1 Support for Struggling Mathematics Learners

Instructional Strategy	Definition	How to Implement
Acceleration (60%–70% of instructional time)	Preteaching of mathematical concepts and a focus on vocabulary	Create a content map of the unit: • Essential questions • Critical concepts and skills • Vocabulary Select the key vocabulary: • Use a variety of strategies. Use graphic organizers: • Link to prior knowledge. • Build concepts prior to the lesson. Put vocabulary in the context of the lesson: • Teach vocabulary twice—first in the intervention, and then again in the context of the lesson. • Consider the prerequisite skills/concepts needed for the upcoming unit.

Instructional Strategy	Definition	How to Implement
Remediation (30%–40% of instructional time)	Skill development	Focus teaching on reasoning skills: • Use the Marilyn Burns program (Do the Math Now). • Use the Math Reasoning Inventory (MRI; free online). Use mathematics skills games: • Integer war • Number operation bingo • Which number is greatest? • Race to see who can order fractions the fastest • Flash card competition Gain additional skills practice from textbooks. Use technology: • ALEKS • Carnegie MATHia • Catchup Math • DreamBox • Khan Academy • LearnZillion • Math Reasoning Inventory (MRI) • Think Through Math
Recovery (10% of instructional time)	Additional time to retake tests or quizzes, or extra time for homework	Take additional time for the following: • Task exploration • Retesting • Homework • Practice

(Adapted from Learning-Focused 2008)

Differentiation by Specific Need

At each specific tier for intervention, students have different needs. Teachers are responsible for all Tier 1 interventions. When a student has additional needs and is identified as a Tier 2 or Tier 3 student, then outside help is needed.

Tier 1 Students

Tier 1 students are generally making progress toward the standards, but at times they may experience temporary or minor difficulties. These students may struggle in only a few of the overall areas of mathematical concepts. They usually benefit from peer work and parental involvement. As Tier 1 students begin to increase their understanding and start succeeding, their confidence level and mathematical disposition grow. When learning difficulties do arise, they should be diagnosed and addressed quickly to ensure that these students continue to succeed and do not fall behind. For these students, teachers may use easier numbers when adding and subtracting positive and negative numbers, so students understand the idea of integer fluency. Once students understand the basic concept, then the teacher can include decimals or fractions into their lessons for deeper understanding.

Tier 2 Students

Tier 2 students may be one or two standard deviations below the mean on standardized tests. These students are struggling in various areas, and these struggles are affecting their overall success in a mathematics classroom. Students who are Tier 2 usually have interventions that focus on academics or behavior. With Tier 2 interventions, the RTI committee finds a way to provide every targeted student required access to the important mathematics for the learning standard in ways that the regular class instruction could not (Kanold et al. 2013). Tier 2 students need additional time outside the regular classroom and additional support to fill in learning gaps.

Tier 3 Students

Tier 3 students are seriously at risk of failing to meet the standards as indicated by their extremely and chronically low performance on one or more measures of formative, summative, and standardized tests. Often some type of in-house student assistance team is analyzing these students to look for overall interventions and solutions. Tier 3 students need the most intense and individualized interventions outside of the regular mathematics class. If the interventions are not successful, these students are usually considered for special education.

English Language Learners

English language learners are learning concepts and language simultaneously. They need to have context added to the language, and this requires purposeful planning by the mathematics teacher. While these students may have acquired social language skills, the language of mathematics is very academic in nature. Students will need ample time to engage in talking about mathematics with their peers. Teachers of English language learners can make using mathematical language a game. The technical language needs to be broken down by the mathematics teacher and made explicit in the mathematics classroom. This is one of the most important keys to success with English language learners, while these students acquire the necessary vocabulary for greater comprehension of the course content.

Above-Grade-Level Students

Many of the objectives covered in a mathematics class are new to students. All students need a firm foundation in the core knowledge of the standards. Even above-grade-level students may not know much of this information before a lesson begins. However, the activities and end products can be adapted to be appropriate to those students' individual levels. High-achieving students can create a different end product in order to demonstrate mastery. Differentiation calls for these students' needs to be met—keep them engaged. For example, a teacher can ask above-grade-level students to design their own experiment to see if the number of hours a student studies affects his or her mathematics grade.

Instructional Intervention Settings

The PLC can choose between many different types of intervention settings depending on the needs of students, assessment results, resources available, and time limitations. These are described in the sections that follow.

Small-Group Instruction

This intervention setting includes, but is not limited to, pull-out programs for struggling students, special education students, and/or English language learners. In small-group instruction, students receive additional instruction on the mathematical concepts with which they have difficulty and more concrete practice of basic mathematical skills and targeted standards.

Mathematics Review Class

This intervention setting is a mathematics class in addition to a student's regular mathematics class. This class includes a small number of students (10 to 12) who need more time with vocabulary and skill development, more time to explore a concept, and additional opportunities to demonstrate mastery of grade-level standards. For example, if students are learning unit rates in their regular mathematics class, the mathematics review class can support this work by building fraction sense, finding equivalent fractions, and working with ratios. This is also a good time to work on vocabulary with these same words and provide thought-provoking problems that include unit rates.

It is important that students perceive this class positively. Working on mind-set (see Chapter 1) will help students experience the benefit of hard work and realize their intelligence is not fixed. As a result of the work they do in a mathematics review class, students will experience more success in their regular mathematics class and gain self-confidence.

Before-School and After-School Programs

In before-school and after-school programs, students receive quiet instructional time away from the distractions of the regular school day. Instead of being pulled out of an activity or their regular classroom, students during this time can receive remedial help or prepare for future class lessons through extra review and practice.

Saturday School

Similar to a before-school or after-school program, Saturday school provides a time for students who normally have weekday outside community commitments, such as sports practice, religious activities, or volunteer opportunities, to receive additional mathematics instruction. Although a somewhat unusual intervention requiring extra funding or volunteer teaching, Saturday school can offer concentrated time for students to focus on learning mathematics. With the right atmosphere, students have been known to look forward to Saturday school, especially if they have been given some choice in attending.

Summer School

A district may offer summer school for students before they begin a new course or as a remedial time for students who need to strengthen their understanding of mathematical skills introduced during the previous year. Because summer school is typically scheduled for four to six weeks, students can improve their skills at a slower pace and in a more focused manner.

Differentiation Strategies

In a mathematics class, essential learning should guide instruction for student understanding. But for some students, understanding the essential learning does not happen in one class period. These students need more time to master a mathematical concept; others might need intense focus on skills that support the essential learning. Not all students learn at the same pace and in the same time. Teachers should meet students where they are ready to learn and offer support where needed.

Figure 7.2 summarizes differentiation strategies teachers can use for students in Tiers 1, 2, and 3.

Figure 7.2 Differentiation Strategies

Strategies for Tier 1 Students	Strategies for Tier 2 Students	Strategies for Tier 3 Students
Use formative, summative, and diagnostic assessments to identify the areas in which these students are not at mastery. These students benefit from pair work in which sometimes a student is the teacher (in the areas in which this student excels) and sometimes the student is the peer learner. Reteach concepts in a different way. Allow small groups to study concepts together; encourage student talk. Ask for parental involvement in keeping these students on task in assignments. Allow partner work for students to check their work and build confidence that they are on the right track. Allow for multiple entry points. Allow partner work for oral rehearsal of solutions. Provide cheat sheets as needed.	Use formative, summative, and diagnostic assessments to identify the areas in which these students are not at mastery. These students benefit from a before-school, mathematics review, or after-school intervention program. Reteach concepts in a different way. Offer extra practice in areas of struggle with study groups or peers. Allow time for peer tutoring. Extend mathematics instruction. Model often, showing them step by step how to solve problems. Allocate extra time for teacher-guided practice. Use activities centered on students' interests.	Use formative, summative, and diagnostic assessments to identify the areas in which these students are not at mastery. These students require extra intervention programs. Consider a before-school or after-school program, or extended mathematics periods to combat the risk of failure. The school's student assistance team can determine if students in this category might need testing for special education needs and an individualized education program (IEP) or modifications.

Figure 7.3 shows differentiation strategies that can be used for English language learners or for above-grade-level students.

Figure 7.3 Additional Differentiation Strategies

Strategies for English Language Learners	Strategies for Above-Grade-Level Students
Allocate extra practice time to apply and use the vocabulary with the concepts.	Offer accelerated processing activities.
Allow more time to simultaneously process the language and the content.	Modify assignments.
Use visual displays, illustrations, and kinesthetic activities.	Design assignments where students are working at Webb's Depth of Knowledge level 3 or 4.
Offer notes that are partially filled in so that students can focus on necessary information.	Assign problems with greater complexity and creativity.
Start with concrete examples and use manipulatives.	Have students research a question related to the mathematics standard.
Plan for oral rehearsal with partners of the academic language behind the mathematical concepts.	Request oral presentations of the concepts, which will benefit all students in the classroom.
Evaluate the use of word problems. Read them aloud and emphasize key words that indicate procedural action.	Provide real-world problems where students need to think creatively about a solution.
Allow for partner work.	Ask students for ideas that will help them grow as mathematicians.
Model think-alouds.	

Grouping Strategies for the Classroom

Chapter 4 discussed multiple ways to group students, as students need to engage with one another to discover and learn mathematics:

- Individuals

- Pairs

- Small groups

- Whole group

Teachers should choose a grouping strategy based on the mathematical task.

The grouping strategies offer opportunities for the teacher to observe what students know about specific mathematical concepts. They also offer

Tier 1 and Tier 2 give students a chance to engage in the mathematics without being singled out.

The sections that follow offer grouping strategies suggestions based on the goal for instruction and ideas for group activities.

Choosing a Grouping Strategy

Figure 7.4 shows some of the most effective grouping strategies based on the teacher's goal for instruction. Please note that the teacher could choose one grouping strategy for a particular goal in the lesson or use multiple strategies to effectively reach a single goal.

Figure 7.4 Grouping Strategy Based on the Goal for Instruction

Goal for Instruction	Grouping Strategy
Deliver lesson instruction to all students.	Whole-group instruction
Check for understanding during the lesson.	Pair, small-group, or individual activity
Have students think about and practice giving an answer to a question before responding to the teacher.	Pair-sharing activity
Have students practice a concept.	Pair, small-group, or individual activity
Offer practice in using the academic language of mathematical vocabulary.	Pair-sharing activity or small-group activity
Have students work in groups while reteaching needs are identified.	Homogenous small groups
Have students teach and learn from each other while practicing a concept.	Heterogeneous small groups
Have students orally apply a concept.	Pair or small-group activity
Use manipulatives.	Pair or small-group activity
Have students apply a concept.	Pair, small-group, or individual activity
Offer self-supporting practice time.	Individual activity
Have students demonstrate mastery of a lesson concept.	Individual assessment

Group Activities

The following are examples of the variety of group activities that engage learners and allow teachers to differentiate the pacing, practice work, and actual instruction during mathematics lessons:

- Cooperative grouping caters to the social side of students' personalities. It also gives them opportunities to practice the academic vocabulary surrounding the concepts. Group activities often boost the confidence and risk-taking that will later affect individual student work. The teacher can say, "Class, you have two minutes to use today's terms, *odd* and *even*, in as many ways possible. Go!"

- Physical games help kinesthetic learners remember the key concepts and get students out of their seats. For example, students can walk a coordinate plane to learn the x and y values of a coordinate pair. Some students might decide to act out the terms *odd* and *even*.

- Whiteboard games or card games challenge students and encourage learning. Playing games as a whole class on the board or overhead projector can serve as a form of informal assessment as the teacher checks for student understanding. These games are fun and educational, and they also help ensure that all students are successful. The card game War is a great game for students to play when they are learning how to count and comparing numbers.

- Large-group games provide a safe environment for lower-level students to take risks. These students are supported by their teammates and can stretch their thinking outside of their comfort zones. Using a gallery walk demonstrating student work on posters is an effective way to get all kids involved.

- A game based on the concept of the teacher "competing" against the class allows students to participate and have fun while trying to beat the teacher—for example, the teacher can play against students in a factor game.

Conclusion

Getting to know students as mathematics learners is indeed an intricate process. Even for veteran teachers, sorting through the complexity of differentiation can be overwhelming. Using the suggestions and guidelines in this chapter and committing to one or two strategies, teachers will begin to see success. Each student has specific, individual needs. Keeping these in mind and planning for purposeful mathematics instruction will help teachers navigate the differentiated classroom.

Reflection

1. How are differentiation strategies for English language learners similar to differentiation strategies for Tier 1 and Tier 2 students?

2. List three ways you can differentiate mathematics instruction for students who are performing above grade level in your class. How can you avoid busywork and engage them in deep, meaningful mathematical work?

3. How can grouping strategies help when differentiating instruction?

Chapter 8

Integrating Mathematics Across the Curriculum

In the past, it might have taken teachers up to two years to successfully navigate and understand state standards documents. Standards were often taught out of context and in a lecture setting, so learning rarely stuck. Now with fewer yet deeper standards to master, students have opportunities to explore mathematics and make connections to real-life situations. According to Steven Leinwand (2000, 85), the author of *Sensible Mathematics: A Guide for School Leaders*,

> "For too long, for too many students, learning mathematics has meant moving from topic to topic, and from chapter to chapter, with little regard to the connections between and among these topics or chapters, and even less attention to the connections between the topics and their application in the world and other disciplines."

When students study a mathematical concept connected to a real event or place, they tend to understand and remember it. Leinwand (2000, 85) also states, "Math being learned in real-world contexts makes learning easier, more enjoyable, and more significant."

As teachers help make connections from the mathematics class to the real world, students begin to see how mathematics is all around them and not just in isolated problems or in a textbook. At this point, integrating mathematics in other content areas becomes feasible. Working either independently or closely within a professional learning community (PLC), teachers can create deep, rich, and coherent integrated units of study. Bear in mind that the process can take a lot of time, collaboration, planning, and flexibility. Teachers need to be careful not to create "filler" activities but instead strive to develop meaningful and thought-provoking tasks. They can start with the standards and write overarching goals of the integrated unit of study. Then they can write one or two essential questions for students to explore, and start working backward when developing lessons or activities. It is helpful for teachers to think of the following questions when planning integrated units of study:

- How much time is needed to finish the integrated unit of study?

- What will the assessment look like?

- What activities require planning?

- Who is involved in planning and teaching the activities?

- What choices can the teacher provide for students as they work?

- How can the integrated unit of study relate to students' lives?

Elementary teachers have an advantage when integrating mathematics into other content areas, since they are usually with one set of students all day or can work with grade-level partners to plan integrated units accordingly. It becomes much more challenging at the secondary level to integrate mathematics into other content areas. Secondary teachers often teach in isolation and focus on depth in their singular discipline, so they may feel that they have no time to devote toward collaboration with other content. Some secondary teachers may also fear dilution of their subject. However, once they experience the insights that students gain in an interdisciplinary unit, including increased depth in their particular subject, secondary teachers become advocates of cross-curricular work. It is important for students, both elementary and secondary, to see that mathematics is found not only in a mathematics class, and to see the connections between mathematics and other content areas.

Rather than working on subjects in isolation from one another, students learn best when they are engaged in inquiries involving academic language and naturally incorporating content from a variety of subject areas (NCTE 1993). It is important for students to understand that education is not a series of compartmentalized subjects that have nothing to do with one another; rather, learning is more like a tapestry, where all subjects are woven together to create a broad scope of understanding that is ultimately most useful when all the strands fit together. This chapter provides suggestions and strategies for teachers to integrate mathematics across the curriculum. It includes information about reaching all learners and broadening students' understanding of mathematical concepts.

Integrating Mathematics and Literacy

Mathematics and literacy are synonymous in mathematics classrooms. Teachers are asking students to read, write, reason, and make inferences about mathematical tasks. The Common Core State Standards for Mathematics (CCSSM) require students to reason abstractly and quantitatively. Literacy and mathematics require students to make sense of problems and construct viable arguments while being able to defend and critique the reasoning of others.

The Language of Mathematics

Students learn to read, interpret, speak, and write the language of mathematics as they tackle concepts. The linguistic element of mathematics varies by culture as well. For example, the way fractions are read in English is different syntactically from the way fractions are read in other languages. English starts with the numerator (a cardinal number) and then the denominator (an ordinal number). There are exceptions: students do not say "$\frac{1}{2}$" as "one second," but rather "one half." In Japanese, to express a fraction, the student first states the denominator and then the numerator. The Japanese would say "four sections of which we are referring to one section," for the English equivalent of "one fourth."

A literacy-rich environment can provide a bridge to new terms and meanings to enable new information to be assimilated into students' existing schemata (NCSM 2014). Mathematics is a technical language that requires intense focus on literacy. Integrating reading and writing in a mathematics classroom is essential for establishing real-world context.

Reading

Reading is inevitable in a mathematics classroom, and it is critical that mathematics teachers clearly explain the purpose of reading to their students and teach reading strategies specific to mathematics. Using research by David Pearson et al. (2007), Stephanie Harvey and Anne Goudvis (2007) summarized strategies thoughtful readers used to construct meaning. They found that readers did the following:

- Made connections from prior understanding and new information

- Asked questions

- Drew inferences

- Noted important and less important information

- Synthesized information

- Monitored understanding

Readers take the written word and construct meaning based on their own thoughts, knowledge, and experiences (Harvey and Goudvis 2007). A student needs to be able to read concept introductions, explanations, word problems, examples, and instructions in a mathematics book or within an activity. It is helpful for mathematics teachers to model and explicitly demonstrate the reading skills required in a mathematics classroom. Partnering with literacy colleagues will help strengthen mathematics teachers' ability to model. Students will benefit from the common language. Delving deeper into literacy, mathematics teachers can even explore genres. As they read through a mathematics problem and model their thinking, mathematics teachers can highlight the important information and identify the question that needs to be answered.

Integrating Fiction and Nonfiction

A teacher may apply books to lessons to help contextualize mathematical concepts. Some of the skills needed for reading are also required for mathematical thinking, including summarizing, sequencing, and finding the main idea. Integrating reading and mathematics can improve students' general language skills and their abilities to communicate and express themselves mathematically (Moyer 2000).

Many students of all ages enjoy reading and listening to stories because fiction text is appealing and interesting. A number of fiction texts directly explain mathematical concepts and pose specific mathematics problems to solve in a fun and exciting way. Additionally, some fiction books present opportunities to teach logic through interesting problems established by the characters. These books often present mathematics in an unintimidating way and enable students to see mathematics through a different lens. *The Greedy Triangle* by Marilyn Burns is an example of such a book, where students can see what happens to a polygon as it acquires one more side and one more angle. Students will learn the names of polygons and see where they can be found in real-life settings.

Reading comprehension skills are essential for academic success throughout all grade levels. Students are often taught strategies for comprehending fiction texts; however, as Lorna Collier (2008) states, "…it's essential that students learn from the earliest grades through high school how to read nonfiction, if they are to survive and thrive in an adult world crammed with information." Nonfiction texts naturally provide real-life situations and information that students can use to understand and apply mathematical concepts as well as reading comprehension strategies. For example, a teacher can have students bring in the classified section of a newspaper. Using the used-car section, students graph the year and price of cars they find. The teacher can then ask students to find the correlation and interpret their findings. Here students are employing multiple literacy strategies and ending by writing about their data.

Writing

Writing is an excellent way for students to reflect about a mathematical concept, consider how they worked as a partner, make a generalization about a concept, and think about next steps. As students discuss and read about mathematical concepts, teachers can also guide them in their mathematical writing development. Figure 8.1 presents a few writing activities well suited to a mathematics class.

Figure 8.1 Writing Activities to Support Mathematical Thinking

Writing Activity	Description
Learning logs	The teacher prompts students to write vocabulary, concept ideas, procedures, formulas, questions, reflections, wishes, understandings, thoughts, and examples in the learning log. Students write daily in the learning log.
Sentence frames	Sentence frames help students write mathematics in the proper context. The teacher scaffolds the proper syntax and vocabulary of important mathematical concepts. This process is especially important for English language learners. Example: The _____ measures _____ inches.
Math journals	In math journals, students have opportunities to explore the big ideas. With this quiet ink time students can explain, reflect, practice, and review mathematical concepts. This is where they expand and internalize their mathematical thinking.

Teachers tend to find out a lot about their students through student writing. Although time constraints are real, it is important for teachers to take time during the class period to allow for students to write and reflect. This is where teachers get to know their students as mathematicians. The teacher can model how to write in a learning log or in a journal. As students observe the teacher write and reflect on his or her own learning, they have a model to guide them. Also, it is important for the teacher to provide feedback to students as they write, so they are challenged to improve their writing skills and produce meaningful work.

Always keeping in mind that the focus is mathematics, teachers can ask students to write the following:

- Their understanding of how to solve problems
- An explanation of how they solved a problem
- Their level of comprehension of a certain mathematical skill
- Predictions for solving problems
- Prior thinking and if it has changed
- Questions they have for the teacher
- How they worked as a group member

- A reflection on perseverance

- Whether they have a growth mind-set or a fixed mind-set

- Goals for the next day in math

- If they had more time, what they would have done differently (if anything)

- A review of the steps or procedures involved in solving problems

- Sample word problems or math-related stories where they can apply the mathematical concepts they are learning

- A reflection on new concepts learned that day

- A self-assessment of how well they feel they did on homework, a quiz, a test, or practice work

- Whether they feel challenged

- How they view themselves as mathematicians

Strengthening the literacy support of the mathematics class will stand teachers in good stead and provide support when they incorporate other content.

Integrating Mathematics and Science

Mathematics and science are disciplines that can often be taught together quite naturally. When teachers use science experiments to practice mathematical skills, students are able to understand the relationship between science and mathematics more completely because they are actively making their own meaning and connections. Learning is reinforced through related concepts, leading to better retention in both content areas. Mathematics and science can be easily connected in the following ways, among others:

- **Measurement:** Science experiments are ideal for applying mathematical measurement concepts to real-life situations. The teacher can give students cylinders of various sizes and ask them to make a prediction about which cylinder holds the most water. Students then fill the cylinders with water and measure how much water each cylinder can hold.

- **Geometry:** The natural world is bursting with lines, angles, and shapes to explore. Taking the geometry lesson into life science is a natural way for students to explore geometric concepts. As students explore different three-dimensional shapes, the teacher can take them outside to find shapes that match each of the specific characteristics.

- **Scientific experiment data:** Often, students have difficulty understanding how the formulas they are learning in mathematics relate to the real world. If they are given a chance to use mathematical formulas during scientific experiments, the formulas become more meaningful. The teacher can ask students to use the circumference of Earth to figure out the radius, the distance to the center of Earth, and the diameter if students were to stick a rod through the center.

- **Patterns:** By exploring patterns that occur in nature, the teacher validates patterns studied in mathematics class. For example, when studying the Fibonacci sequence, students can physically count the number of spirals on pinecones that align with the sequence (1, 1, 2, 3, 5, 8, 13, etc.).

- **Algebraic formulas:** Some students struggle with algebraic concepts. If they are able to see algebraic formulas "in action" while conducting experiments, they will better understand the concepts. Students can learn the formula for the Pythagorean theorem, but when they can actually build the perfect squares off of the triangle, they see the connection of the perfect squares and the triangle, as shown in Figure 8.2.

Figure 8.2 The Pythagorean Theorem in Action

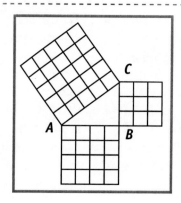

- **Probability:** Both mathematics and science explore the nature of probability. As students conduct experiments, they predict, test, and record the results, which offers them real-world experience with the concept of probability. For example, the teacher can put a blue marble, a red marble, and a yellow marble in a hat and ask students to find the probability of drawing the blue marble. Would it be the same as drawing the red one? Students can set up an experiment to test their theory.

As demonstrated in this section, science is a natural fit for integration with mathematics; however, other subjects can be integrated as well.

Integrating Mathematics and Social Studies

Teachers can integrate mathematical concepts with social studies and history in a number of ways. In social studies classes, students connect with the past, learn about places and people, and apply this learning to the present. Students can use mathematical methods to analyze historical and social patterns. When teachers successfully integrate different content-area concepts, students can more successfully retain information and make connections within their own learning.

It is easy for elementary teachers to integrate mathematics and social studies. For example, asking students to explore mathematicians in a history class or having students use a timeline to study chronological events are a couple ways teachers can integrate social studies and mathematics at the elementary level. It is possible for secondary mathematics teachers to work closely with social studies teachers, but it will require time and collaboration. The end product could be beneficial for students in both classes if the overarching goal connects content standards in mathematics and social studies.

The following sections illustrate strategies and problems where social studies and mathematics are used in unison. As students take time to explore history, geography, cultures, and human issues, a real-world connection is made. By integrating the two curricula, students will investigate interrelated areas and take ownership of their learning.

Word Problems

The most obvious and perhaps easiest way to integrate social studies with mathematics is to create word problems that combine social studies topics with the mathematical concepts students are learning. To make sense of mathematical concepts, students often need to apply them to essential or open-ended questions—for example, "If the Great Wall of China were built to protect the Chinese empire against adversaries, how many people do you believe would need to be stationed in watchtowers? Find out how long the Great Wall of China is and explain whether your prediction is reasonable." Teachers should be sure to use word problems that fit within accepted standards to avoid wasted time.

Connections to Historical Dates

It is often difficult for students to fully understand the concept of the past. Years and dates have little concrete meaning, especially for young children, who are still developmentally processing the ideas of past, present, and future. Students of all ages need multiple opportunities to process information about different eras in order to begin to understand the idea of history. Mathematics can help clarify these concepts, as shown in the following example.

Example: The Japanese attacked Pearl Harbor on December 7, 1941. If a U.S. Navy junior officer was 25 years old on that day and survived the attack, how old is he now?

Solution: (Varies depending on the current year.)

Connections to Historical Data

As students study places and people in social studies, they can use mathematics to connect these places and people to their own lives. They can compare the size of ancient civilizations to the size of their present state (e.g., Arizona or South Dakota) and use census data to compare the number of people living in one area to the number of people living in their city or town. They can also use maps to calculate and measure differences between locations, as the next example demonstrates.

Example: On the map, there are 1,300 miles between Boston and Charleston using the King's Highway, an early American trail. It took almost two months for settlers to walk between those places because they could travel only 20–25 miles per day with their wagons. If you were to walk 25 miles from your city, where could you possibly end up?

Solution: (Varies depending on the location of students.)

Integrating Mathematics and Technology

Mathematics and technology are naturally associated, but they have not always been effectively integrated in the classroom. Students may have been allowed to play computer games that incorporate mathematics drills but then forbidden to "cheat" on activities or tests by using calculators. Today's educators understand that technology, such as computers and calculators, can improve students' education (Dean and Florian 2001). Rather than using computers as drill-practice machines, teachers now encourage students to use spreadsheets, online applications, and more to explore, apply, and display mathematical learning. Students can receive online mathematics tutoring, instruction, and homework help from a number of websites. Computers and calculators are tools that students use to research, gather data, organize their notes on mathematical concepts, and analyze real-world problems and situations.

The use of technology can provide students with opportunities to develop and use their mathematical higher-level thinking skills to solve problems that are relevant to their daily lives. The sections that follow show the integration of technology and mathematics. Technology is ever changing and it is up to the teacher to find the essential learning of the lesson and then see if technology can be integrated within it. Teachers should not use technology for its own sake, but instead to enhance a lesson so students have an opportunity to gain a deeper understanding of the mathematics.

Differentiation Within Technology-Embedded Mathematics

It is completely normal to find a range of mathematical and technical competencies in any group of students. When teachers incorporate technology into mathematics lessons, they have various options for differentiating instruction to meet the needs of all students:

- Teachers can use varied grouping strategies when assigning group mathematics projects. In a center, for example, students may interact with a whiteboard.

- Teachers should vary the level of support given to students as they use computers and calculators to solve problems and display data. If the essential learning is to find the surface area of a rectangular pyramid, for example, it is okay for students to use a calculator to find the values. However, if students are learning addition with two-digit numbers, this would not be a good use of calculators.

- Teachers should be flexible when setting time limits for work that requires technology tools.

- Teachers can allow students to use extra technology tools for scaffolding purposes when necessary. Sometimes allowing students to play mathematics skill games on the computer is a good use of time for remediation.

- Teachers can use multiple technological representations of mathematical concepts. For example, students can graph a linear equation by hand and then check their work on a graphing calculator.

Using Spreadsheets in a Mathematics Classroom

Spreadsheet programs such as Microsoft Excel can be valuable resources with numerous uses in a mathematics classroom. This technology allows students to organize and manipulate data in cells within columns and rows. Spreadsheets offer many ways to display and apply formula functions to numbers. While teachers of all subjects may use spreadsheets to keep records of grades, classroom budgets, attendance, and checklists, spreadsheets are also quite useful for mathematics instruction.

Students can use a spreadsheet program to do the following:

- Manipulate data related to classroom studies
- Build analytical, mathematical, interpretive, and technical skills
- Compute mathematical formulas
- Show mathematical patterns
- Create graphs that show algebra and trigonometry relations and functions

The sections that follow cover ways students can use spreadsheets in the mathematics classroom to create, read, and interpret graph information; manipulate data; and research products.

Creating, Reading, and Interpreting Graph Information

State and district standards may require students to create and interpret different types of graphs. Students can use spreadsheets to predict changes in numbers, build various types of graphs, and compare and manipulate many kinds of data. Spreadsheets can also be used to study a wide range of financial topics. For example, students can use spreadsheets to create and manage a household budget, and they can even display negative balances in an alternate color, such as red.

Formulaic Manipulation of Data

With a spreadsheet, students can learn about, apply, and experiment with many different orders of operation, as follows:

- Adding, subtracting, dividing, and multiplying numbers
- Displaying a value as currency, percent, or decimal
- Calculating means and averages
- Creating calendar formulas and conversions
- Applying arithmetic, mathematical, or statistical functions to data sets
- Calculating running balances
- Representing in graphical or numerical form
- Establishing relationships
- Converting measurements

Middle school teachers can ask students to construct a research project in which they record student heart rates after running for one minute. They must find students in sixth, seventh, and eighth grades, and there must be three trials for data collection. Teachers can have students input their data in a spreadsheet and find the average heartbeats for students in the sixth, seventh, and eighth grades. Then, students can to do the same for each of the three trials. Finally, students can interpret the data collected.

Research Products

Students can use spreadsheets to display information that they have researched. For example, they might research nutritional recommendations, create charts to display the information, conduct surveys of the actual nutritional intakes of their classmates, and create comparison charts in the spreadsheet program. This example combines the subjects of health, literacy, science, and technology to enhance mathematical learning.

Students can also use spreadsheets to organize scientific and mathematical data in the same way. If students are studying temperatures, they can record hourly temperatures. After entering the data in a spreadsheet, they can graph it and analyze the information.

Using Apps in a Mathematics Classroom

Apps, short for *applications*, are software-component programs that perform precise functions on smartphones, tablets, web pages, and desktop computers. Apps can be used as tools to help enrich mathematics instruction and also as resources for students if they need help outside of the classroom. An app is usually designed to perform a single action, and it has built-in restrictions that prevent it from harming the user's computer or electronic device.

Many educational websites provide free apps for download. Some apps are simply for fun, while others have more practical, academic applications. Some websites provide instructions on selecting, installing, and using apps. Teachers should take the time to download and use an app to determine whether it is a useful resource or tool before introducing it to students.

To use apps in a mathematics classroom, students ideally have access to computers at school. While it is not always feasible to have a classroom set of computers, teachers may be able to sign up for a computer lab or have access to a mobile computer lab. If a school is deficient in technology resources, teachers should advocate for their students. Access to various websites offering app tools will help students process and learn mathematical concepts. Some of these sites may have virtual manipulative apps. A teacher and students might find maneuverable algebra tiles, logic puzzles, virtual money for making change, and blank bar graphs, line graphs, and pie graphs that allow students to enter and display data. These apps assist differentiation in the mathematics classroom because students can use the apps that best match the skills they are developing and progress at their own speed.

Figure 8.3 lists a few apps that can be used for elementary, middle school, and high school levels of mathematics as well as some possibilities for their use in a classroom.

Figure 8.3 Apps for the Mathematics Classroom

App	URL	Description
Desmos	https://www.desmos.com	Graphing calculator
DragonBox	http://www.dragonboxapp.com	Educational game for learning algebra
Edmodo	https://www.edmodo.com	Social learning platform
Educreations	http://www.educreations.com	Recordable whiteboard tool to teach or learn mathematics lessons
Khan Academy	https://www.khanacademy.org	Mathematics instruction, including challenges, assessments, and videos
LearnZillion	http://learnzillion.com	Mathematics lessons and tutorials
Motion Math	http://motionmathgames.com	Interactive mathematical facts games
National Library of Virtual Manipulatives	http://nlvm.usu.edu/en/nav/vlibrary.html	Computer-based mathematical manipulatives and interactive learning tools for K–12
Quizlet	http://quizlet.com	Mathematics flash cards
ShowMe	http://www.showme.com	Recordable whiteboard tool to teach or learn mathematics lessons
Skype	http://www.skype.com/en/	Online video and voice calling, file sharing, and instant messaging tool that enables students to visit classrooms around the world
YouTube	http://www.youtube.com or http://www.youtube.com/Teachers	Videos on a variety of topics, including mathematics instruction

Using Graphing Calculators in a Mathematics Classroom

The graphing calculator is an important tool in all areas of mathematics. Graphing calculators can help elementary through high school students use higher-level thinking and apply mathematical concepts to specific problems. Teachers can show students how to use a graphing calculator as a tool for mathematical-concept comprehension and application. If some students are already comfortable using a graphing calculator, it is beneficial to have an "expert" in each group. Students learn best from each other.

A study by Douglas Grouws and Kristin Cebulla (2000) notes that "teachers ask more high-level questions when calculators are present, and students become more actively involved through asking questions, conjecturing, and exploring when they use calculators." If the essential learning of the mathematics requires deep thinking, the calculator is used as a tool to get to that level of thinking. According to the National Council of Supervisors of Mathematics (2014), "Technology-enhanced learning environments help students make real-life connection to mathematical concepts. These digital devices become tools for collaborative, creative learning to enhance engagement and motivation of students in mathematical classrooms."

The next sections cover additional information on the practical use of calculators in the mathematics classroom. As students gain more experience with using a calculator, they will see the value of the tool and how it can help them in their mathematical understanding.

Classroom Management of Calculator Use

Calculators can allow teachers to spend even more time developing mathematical understanding, reasoning, number sense, and application. Here are a few helpful hints to remember when working with calculators:

- Teachers can number all the calculators before distributing them to students and then assign each student a calculator number. This way, teachers can easily keep track of when calculators are returned.

- All class sets of calculators should be distinctly marked with bright paint or permanent marker on the outside cover so that they are easily identifiable by teachers, administrators, and other students if they are taken out of the classroom.

- The calculators should be stored in plastic shoe boxes or in an over-the-door shoe rack with the corresponding numbers printed on the slots.

- Sufficient class time should be allotted for the organized retrieval and return of the calculators. If students will need the graphing calculators at the beginning of the class period, instructions should be displayed as students arrive.

Teaching Graphing Calculator Skills

Students can learn, reinforce, and review mathematical concepts using graphing calculators. Practice with the calculators will solidify students' understanding of concepts such as number sense, algebraic thinking, data analysis, spatial reasoning, problem solving, and units of measurement (Grouws and Cebulla 2000). Teachers can integrate graphing calculators into their lessons with this type of practice in mind.

Here are some strategies for using graphing calculators in class:

- The teacher should give students time to ask questions and explore the calculator functions.

- To teach a skill, the teacher should request that students locate the keys and functions on the calculator. Students can do this in pairs.

- The teacher should familiarize students with the menus and screens that they will use often.

- If multiple steps are needed to complete an activity, the teacher can list the steps on a poster display, whiteboard, or Smartboard so that students can reference the steps during the lesson.

- The teacher can designate students who are more familiar with calculator use to assist those who are not.

Building Conceptual Understanding with Graphing Calculators

Graphing calculators are capable of providing multiple representations of mathematical concepts. By building tables, tracing along curves, and zooming in on critical points, students may be able to process information in a more varied and meaningful way (Smith 1998).

Graphing calculators can build on conceptual understanding by allowing students to practice numerous representations of concepts and experiences in a way that is not possible using paper and pencil alone. As a result of these methods, teachers are able to engage students more effectively by addressing different learning styles and developing understanding that leads to higher-level thinking. Teachers do not often associate the use of graphing calculators with the conceptual process. Graphing calculator lessons can engage students in building conceptual understanding while giving the practice necessary for procedural proficiency in calculator use. As students move through each phase of learning, they are exposed to a concept or skill numerous times when teachers use the following approach in a lesson:

- **I do:** The teacher models the complete procedure using a display product such as a computer program, webcam, or projector.

- **We do:** While the teacher repeats each step, students engage in the step as a class.

- **You do:** Students practice the complete procedure independently, asking their group members for assistance if needed.

To familiarize students with a graphing calculator, teachers can find a simple task for students to complete, such as graphing a line or plotting points. It is helpful if teachers use the "I do, We do, You do" approach just described.

To graph a line (or any function) using a TI-84, follow these steps:

1. Find the *Y=* button that is located at the top left of the TI calculator. Press it and the screen should look like this:

```
Plot 1   Plot 2   Plot 3
\Y1=
\Y2=
\Y3=
\Y4=
\Y5=
\Y6=
\Y7=
```

2. Enter an equation, such as $y = 5x + 9$.

3. Press the *Graph* button and observe the graph of the equation on the coordinate plane.

4. If the line is not on the screen, try pressing the *Zoom* button, then press *ZoomFit* or 0 (both do the same thing), and try pressing the *Graph* button again.

5. Use the *Trace* button to display different "if x equals, then y equals" values.

After the teacher has modeled this task, he or she walks students through using another problem. Finally, students graph functions on their own or with their group.

It is important to use graphing calculators regularly. Students tend to forget how to use them if they are not often featured in lessons. A graphing calculator is not the be-all and end-all; it is very important, but teachers should investigate other electronic possibilities. For example, the application Desmos is another way for students to input, graph, and analyze data. Other programs and apps are available, and it is up to teachers to find the technology that best meets students' needs.

Conclusion

Connecting mathematics with other content areas provides real-world situations in which students can explore authentic problems. Mathematical concepts and procedures come to life as students investigate their own methods for finding solutions. According to Charles Schwahn and Beatrice McGarvey (2012, 92), "If schools are for the purpose of getting learners ready for life, for 'empowering all learners to succeed in a rapidly changing world,' then school and real life have to meet in learning activities that take place in a real world context." Technology-rich classrooms help students explore, develop meaning, and think deeply in all subjects integrated with mathematics to "make learning easier, more enjoyable, and more significant" (Leinwand 2000, 85).

Reflection

1. Is it important to integrate mathematics across the content areas? Defend your thinking.

2. Consider ways you can encourage collaboration with your content colleagues. How will doing so enhance your mathematics classroom?

3. Describe how students can use a mathematics journal in each of the following content areas: reading, writing, social studies, science, and mathematics.

4. Find an app that you can integrate into a mathematics lesson you will teach this year. Write about how you will use the app in that lesson.

New Teacher Support

Teaching is an art. This appendix offers general suggestions for all teachers, but especially those just entering the profession. Some of these recommendations have applications in the mathematics classroom, but the emphasis is on how to excel as a teacher overall.

Professional Support Options

With a high percentage of new teachers not returning to the profession within their first five years, it is imperative that all educators support new teachers. Most school districts offer a new teacher induction program or mentor program for beginning teachers. If teachers are not provided with a mentor or induction program, they can join an online group for new teachers or ask a teacher who frequently stops by to check in to be a mentor. It is especially helpful for a new teacher to find a veteran colleague in the same discipline who can help him or her navigate the early years.

The key to a first-year teacher's success is being surrounded by supportive people. New teachers should strive to spend time with people who will listen, offer advice, push for excellence, and celebrate successes. Novice teachers should try to connect with other like-minded, positive teachers who put students' needs first.

Collaboration is essential when planning for instruction, figuring out classroom management, facilitating, assessing, and differentiating. Working cooperatively can help new teachers find support, commiserate on setbacks, and celebrate successes. If offered the assistance of a coach or mentor, a new teacher should take advantage of that offer. Teaching is very complex and requires purposeful planning and ample time for reflection, which a coach or mentor can assist with. Teachers should not go through their first year alone.

Finding a Mentor

Teaching is not done in isolation anymore. With professional learning communities (PLCs) on the rise and collaboration a natural occurrence in most schools, teachers do not have to figure out everything on their own. Mentoring can happen in a couple of ways. Figure A.1 shows how to get through the logistics of a school day. It is helpful for a new teacher to have someone more experienced to turn to who can give the lowdown on how the school operates or serve as a go-to person when questions arise.

Figure A.1 First-Year Teacher's Survival Guide

Things to Know	Things to Do
Building procedures	• Send for copies/make copies • Routes for natural disasters (fire, tornado, etc.) • Technology checkout • Equipment checkout • How to secure the school when students leave • Supervision
School policies	• Attendance and tardy policies • Reporting teacher absences • Arranging for a substitute • Bus schedules • Bell schedules • Dress code (teacher and student) • Office referrals • Student welfare • Nurse referrals • Discipline • Homework • Grading • Food policies

Things to Know	Things to Do
Resources	• Office forms • Lesson plans • Grade book (paper and online) • Student and staff schedules • Textbooks, workbooks, and teacher editions • Computer labs or laptop carts • Student code of conduct • Employee handbook • Professional reading materials • Office supplies • Office and janitorial personnel
Information for students	• Cumulative files • Schedules • Test score data • Supply list • Discipline plan • Emergency forms • Syllabus • Class norms • Parent letter

Another way a mentor can help a novice teacher is by planning and reflecting. A mentor teacher and a new teacher can plan a lesson together, teach it separately, and then reflect on what went well and what to do differently next time. The mentor teacher helps provide the "what" and "how" of teaching. It is up to the new teacher to listen and implement these suggestions, ask questions, and offer ideas as well.

It is also beneficial for the veteran teacher and novice teacher to look at student work together. If the teachers give a formative assessment on area and circumference of circles, for example, they should correct each other's tests so they are consistent when they grade. When teachers collaboratively examine student work, they are focusing on student understanding and what it should look like. Content conversations arise, which lead to idea sharing, which leads to differentiation planning. It is worthwhile for teachers to make time to have these conversations, not just at the beginning of the school year but regularly throughout the year. As teachers continuously look at student data, it is more likely that they will come to view the progress of *all* students, rather than just the students in their own classroom, as their responsibility.

The mentor teacher can also provide an open-door policy, so the new teacher can come in at any time, observe the mentor teacher teach, and reflect with the mentor teacher afterward. If the new teacher is confused about how to teach a mathematical concept or how to structure student collaboration, this method also lends itself to meeting those needs. This approach demands a high level of trust between colleagues, so it is crucial for the new teacher to offer objective observations and questions rather than subjective observations and statements. The novice teacher is there not to judge, but to learn, and he or she should consider asking open-ended questions.

Finding a Coach

Many school districts offer the use of an instructional coach for new teachers. Per Jim Knight (2009, 2), "What coaching offers is authentic learning that provides differentiated support for professional learning." Coaches can serve in the same capacity as a mentor, but the focus can range anywhere from classroom management, to instruction, to content. Coaches usually support teachers in areas of their choice. An effective coach believes teachers have it within themselves to find a solution. If teachers do not have the experience or knowledge to move forward, the coach switches gears and mentors the teacher.

According to Diana Sweeney (2011, 145), "The coach builds a trusting, respectful, and collegial relationship to serve as the foundation for the coaching work." The teacher trusts that the coach is there to help him or her succeed. The coach trusts the teacher is willing to do everything in his or her power to see students succeed. Coaches can be used as a sounding board to help sort through a teacher's thinking process. Effective coaches stretch and support teachers' thinking and actions as they attempt to develop and improve in their teaching ability.

Coaches can also be seen as job-embedded professional development. Coaches ask teachers to reflect on their practice, study best practices in research, and apply their new learning. This professional development is ongoing and consistent. Coaches encourage teachers to set goals. These goals may include instruction, content, group dynamics, classroom management, student learning, differentiation, and questioning and feedback. The coach then collects and analyzes the data and goes over the data with the teacher. The coach and teacher plan next steps based on the evidence from the data collected. Data can be collected through informal conversations, formal classroom observations, reflecting conversations, or video recordings. Data collection helps teachers see what it looks like when they teach or their students learn (Knight 2014). The coaching cycle continues until the data confirms the teacher's efficacy has strengthened so he or she can work to attain the goal independently from the coach.

As part of job-embedded professional development, the coach may also invite a new teacher to participate in informal observations. Informal observations are when the coach takes the new teacher into another teacher's classroom and asks him or her to see, hear, and notice a specific focus. Elena Aguilar (2013, 107) states, "It's critical that your observations are planned, structured, and focused." After the observation, the coach facilitates a meeting between the host teacher and the new teacher. At this time, the new teacher describes his or her observations and asks the host teacher questions about his or her craft. The coach should provide multiple opportunities for new teachers to observe others.

Tips and Advice for New Teachers

With or without a mentor or coach, a new teacher can apply the following tips and advice to become the most effective teacher possible. Teachers should remember that although they cannot do it all, they can be gentle with themselves, take care of themselves, learn from their mistakes, and show up every day ready to do their best.

Know Your Content

It is important for mathematics teachers to master their discipline. They need to know how to arrive at a solution: the right way, the wrong way, the long way, and the shortcut. Mathematics teachers should choose tasks that are conceptual and meaningful for students. The more teachers know about their content, the less often they are caught off guard by unexpected solutions or solution paths. Teachers also need to know what students have learned before their grade level and where the mathematics is headed in the next grade level. The more content knowledge a mathematics teacher has, the more flexible he or she can be with student learning.

Understand How Students Learn

Students need choice, freedom, autonomy, and collaboration when diving into a rich mathematical task. Students are social beings, and if teachers try to suppress student interaction, power struggles occur and no learning happens. Teachers should plan for students to interact on a daily basis and make sure to provide a task that is relevant to the learners' needs, related to the real world, and challenging, with multiple entry points. Mathematical tasks must be worthwhile problems that require students to focus on explanations and justifications while engaging in productive discourse.

Be a Teacher, Not a Friend

Teachers are professionals. There is a fine line between getting to know students and being their friend, and teachers should not cross it. If teachers try to be too friendly with students, the students will not take the teacher seriously when asked to engage in mathematics. Teachers need to get to know their students on a personal level but respect the student–teacher relationship. Such teachers might not be liked, but they will be respected and their subject will be taken seriously.

Demonstrate and Teach Respect

This section's heading sums up a solution to most classroom management problems. Teachers should talk about respect with students and have students think about what respect means, what respect looks like, how it feels to be treated with respect, and ways to treat others with respect. They should not mention respect once and think that students understand what it means, looks like, and sounds like. At all times teachers should model respect for students, even when it seems impossible. They should regularly offer students opportunities to reflect on their own actions and discuss how they were respectful or if they need some help being respectful.

Plan, Plan, Plan, and Then Be Flexible

Intense planning should happen with every mathematics lesson. Teachers should think about what the overarching standards are, how they break down into learning targets, what students will demonstrate when they are successful, and how to differentiate for students who still do not understand. And then teachers should be flexible in their planning. It is rare for all learning to happen as planned on any given day. When teachers are flexible in their thinking and their actions, a relaxed, natural flow in student learning is the result.

Freedom of Speech: Yes! Shout-outs: No!

Shout-outs allow for invisible students to stay invisible. Students who always know the answer will be the first to shout out, while students who take time to think and process may disengage as learners if they are not given opportunities to process and then participate. Teachers can use the alternative strategies described in Chapter 1 to reduce shout-outs.

Dress the Part

Part of the struggle of being a professional is dressing like a professional. It can be a difficult but necessary transition for young teachers who may be used to a more casual style. Students tend to treat teachers who are well groomed and dressed nicely with more respect than teachers who dress more casually. The teacher is the professional and needs to dress the part, setting an example for students.

Pay Attention to the Ratio of Teacher-Student Talk

Most first-year teachers want to explain everything to students, but they would do well to slow down, take a breath, and stop talking. Teachers need to trust that students can and will learn mathematics. If teachers talk too much, they tend to give away the punch line, and the exploratory part of the lesson is wasted. Teachers are the masters of their own content; it is hard to break away from the "Listen to me—I know best!" syndrome. It is helpful to have a coach record who is doing most of the talking in the classroom. Raw data is hard to argue with and can help a teacher keep the focus on student discourse.

Remain Calm and Firm

No one wins in a power struggle. To avoid power struggles, a teacher can offer students choices—for example, "You can start working on the mathematics now or in five minutes"—and then let them make the choice. By sharing control of the situation, the teacher is valuing students as people. When a student does or says something inappropriate, it is best not to engage with the student in front of the entire class. The teacher must remain calm and remember that he or she is the adult in the classroom; how the teacher reacts speaks volumes to the rest of the students. It is not helpful if the teacher raises his or her voice, reacts negatively, or has to have the last word in a power struggle with a student. The teacher should calmly yet firmly address the situation, state the desired outcomes, and then follow through.

Be Honest with Students

There will be times in a teacher's career when he or she will not know an answer. That is okay! It is best to be honest with students. If the teacher does not know an answer, it truly is a teachable moment. The teacher should tell students he or she will try to find the answer to the question. This shows the teacher is human and does not know everything, and also that the teacher is willing to go the extra mile to learn something new.

Be Proactive with Parents

Teachers should make contact with parents immediately, even if it is just a handwritten note with some background information. Parents want to know with whom they are entrusting their children for the day. After this initial contact, teachers should make a point to call the family of each and every student directly throughout the year.

Notes sent home do not always have to be academic in nature. It is also easy to call a student's home and talk to the parents about what the student is not doing well or right, but it takes time and effort to find something positive to call home about.

Keep Up on the Latest Research

Teachers need to stay up to date with the current research and best practices in teaching. New teachers can ask veteran teachers which books helped them throughout their teaching experience, or ask the administrators or instructional coaches for reading suggestions. Teachers can also get a subscription to a teaching magazine or journal. It is very helpful if teachers can form a study group to read and discuss academic literature. Study groups are a time when many questions get answered and new action plans come to life.

Nurture All Kids

Teachers should love kids—all kids. This is usually the primary reason they became educators. Kids are funny, with their personalities developing more every day. Teachers should take the time to get to know their students and their interests. They should keep an open mind about their students, even on discouraging days, and be prepared to work on their own biases and stay open to loving kids.

Celebrate Achievements

Many teachers do not take the time to celebrate achievements. It is easy to get caught up in all of the shortcomings of a given day:

- "I didn't grade these papers."
- "I handled this situation so poorly."
- "My lesson was a flop!"

There is always something more to do, so teachers should take a minute every day and find something to celebrate:

- "Lorelai figured out what fractions are!"
- "I put up a new bulletin board."
- "All students participated in the summary."
- "I graded my formative assessment today."
- "I am alive at the end of the day!"

No matter how small or insignificant these may seem, they are accomplishments and they should be celebrated.

Take Time to Reflect

Teachers ask students to reflect every day to build a strong understanding of different mathematical concepts or skills. Teachers also need to reflect to find areas of strength and areas needing improvement. Teachers should take a few minutes after every lesson to answer these questions:

- What went well?
- What could I do differently next time?

Hopefully in time, reflection becomes a habit. The more time teachers take to reflect, the more their teaching craft develops. If feasible, teachers should use a coach or mentor and have reflecting conversations about daily lessons, pedagogy, values, and beliefs.

Cultivate a Support System

It is important for teachers to develop a support system within the school and outside of it as well. They can find other first-year teachers who understand the ups and downs a new teacher experiences, other teachers who understand the frustrations and triumphs of teaching, and a friend or family member who can offer an ear to listen. Sometimes it is nice to vent, but teachers should try to do it quickly and move on, as it is easy to get stuck in a downward spiral of whining. Also, new teachers can take time to listen to others. Sometimes teachers get so consumed in their own lives, they forget about all the other people in the world who support them.

Find an Outlet to Relieve Stress

Many first-year teachers spend most of their waking moments at school, which is one reason for the high burnout rate among new teachers. Teachers should establish a leave time and stick to it. They can take time to read a book, mountain bike, hike, swim, run, play chess, spend time with family, or engage in any other hobby. It is healthy to invest time outside of school hours in other interests.

Do Not Let Your Job Define You

Again, for new teachers, it is easy to spend every waking minute working on school-related tasks, but they should not. Teachers need to make sure they have a school-free weekend every once in a while—no grading papers, lesson planning, or answering work-related emails. Teachers need to balance time between work and family and be present for the people in their lives. It will benefit them in the long run to honor this time.

Ask for Help

Someone once said that teaching is not rocket science. Most teachers tend to disagree—it is hard! When the going gets tough, teachers should ask for help. All teachers have been the "new kid on the block" and remember what it was like. No first-year teacher has it all figured out. Actually, no veteran teacher has it all figured out, either. Should a teacher receive a rebuff or a judgmental response, he or she should simply move on and ask another teacher. (And vow to never respond that way to a new teacher.)

PLCs are breaking down barriers, allowing teachers to collaborate together to figure out this "rocket science." New teachers should be brave and ask for help when questions and problems arise, using the knowledge and expertise of administrators, coaches, and other colleagues. These more experienced educators have likely found themselves in similar situations and can offer sound advice.

Final Note

As teachers take the time to read and apply the research–based strategies and ideas provided in this book, they will find their own style as an educator. New teachers should make a commitment to continually develop as learners in the field of mathematics teaching, apply the strategies recommended here, and create some of their own. Taking chances and trying new techniques will help teachers expand their knowledge and commitment to their students. Richard DuFour, Rebecca DuFour, Robert Eaker, and Thomas Many (2010) said it best: "We learn best by doing."

As a final piece of advice, teachers should take time each and every day to reflect. Every time a teacher reflects, true understanding of teaching strategies and practices emerges, and true understanding of the self and the self as teacher is revealed.

Reflection

1. Who can you approach to be a mentor or coach? In what ways can this person help you in your first year of teaching? How will you know they have helped?

2. What tips and advice will keep you focused in your upcoming year?

3. Pick one tip or piece of advice from this chapter and create an action plan on how you will incorporate it in your teaching.

4. What advice would you give a first-year teacher? How does your advice change as you progress during your first year of teaching?

Parent/Guardian Letters

Parent Guardian Letter 1

Dear Parent/Guardian,

I am excited to have your child in _____ grade!

_____

Some of the big math ideas for this grade are:

_____

_____

_____

In this class, your child will work on mathematical concepts guided by state or district standards. Your child will study these concepts by

- using hands-on learning tools
- focusing on key mathematical ideas
- investigating problems
- using higher-level thinking involving different solution paths

Your child will be working individually, in collaborative groups, or with me. I will assess all students frequently to determine whether concepts need to be introduced, reviewed, or retaught.

Please encourage your child as he or she learns the foundational concepts. Please be available to play math games with your child, provide a quiet place to work on homework, and to continually ask your child to justify how they got their answer because there are multiply ways to get to one solution!

Feel free to inquire about your child's progress or let me know of any problems as they arise.

Please read the attached norms regarding our classroom. The students and I have collectively developed these beliefs. We have three rules regarding behavior:

1. Be respectful to self and others.

2. Be responsible to self and others.

3. Take care of personal/school materials.

If you have any questions, please feel free to contact me at any time.

Please take a minute to read through this letter with your child, and sign and return it.

Sincerely,

_____

Teacher Signature

_____

Parent/Guardian Signature and Student Name

Parent Guardian Letter 2

Dear Parent/Guardian,

Your child is beginning a new mathematics course. The course is titled

_____

The program, texts, and resources that I will use are

_____

_____

_____

In this class, your child will work on mathematical concepts guided by state or district standards. Your child will study these concepts by:

- using hands-on learning tools;
- focusing on key mathematical ideas;
- investigating problems; and
- using higher-level thinking involving different solution paths.

Students will be working individually or in collaborative groups. I will assess all students frequently to determine whether concepts need to be introduced, reviewed, or re-taught.

Please encourage your child as he or she learns the foundational concepts, and help him or her establish a quiet place to study and finish any necessary homework. If you are not able to help your child with mathematics problems, please help your child write down which areas he or she is struggling with so that I can assist as needed.

Feel free to inquire about your child's progress or let me know of any problems as they arise.

Please read the attached norms regarding our classroom. The students and I have collectively developed these beliefs. We have three rules regarding behavior:

1. Be respectful to self and others.

2. Be responsible to self and others.

3. Take care of personal/school materials.

If you have any questions, please feel free to contact me at any time.

Please take a minute to read through this letter with your child, and sign and return it.

Sincerely,

_____

Teacher Signature

_____

Parent/Guardian Signature and Student Name

Student Letters

Student Letter 1

Dear Student,

Welcome to _____ grade! I am really excited you are here.

_____

The big math ideas for _____ grade are:

_____

_____

_____

I am here to help you become a great mathematician. It is important that you come to school each day ready as a learner. That means you should

- be curious, open-minded, and thoughtful
- be accepting and patient of others (not everyone learns the same way you do)
- be respectful of the risks people take (listen carefully to others and do not interrupt)
- be willing to take risks (share ideas and other ways of finding solutions)
- be responsible for your actions
- help others be responsible for their actions

As a mathematician, it is very important that you participate in the lessons. This will help you grow as a learner.

Please let me know if you are struggling and need more practice or time. We will work together. We will have fun and learn math at the same time!

The rules our class made are:

1.

2.

3.

4.

5.

Sincerely,

_____

Teacher Signature

_____

Parent/Guardian Signature and Student Name

Student Letter 2

Dear Student,

You are starting a new mathematics course. The course is titled

_____

The books and resources that you will use are

_____

_____

_____

This course will help you develop as a mathematician. It is vital that you take an active, meaning-making stance as a learner. You will be required to do the following:

- Take a learner's stance (be curious, open-minded, and thoughtful)
- Be accepting and patient of others (not everyone learns the same way you do)
- Be respectful of the risks people take (listen carefully to others, do not interrupt, and be aware of body language)
- Be willing to take risks (share ideas, add to brainstorming, and suggest alternative strategies)
- Be responsible for your actions
- Help others be responsible for their actions

As a mathematician, it is essential that you participate in the lessons in order to grow and develop as a learner.

You are responsible for letting me know if you are struggling with some of the concepts and need more practice or time. We will work together to understand, to make sure you are successful.

The norms our class developed are as follows:

1.

2.

3.

4.

5.

If you are not able to follow these norms, what are some alternative solutions we can agree upon? _____

Sincerely,

_____

   Teacher signature

_____

   Student signature

References Cited

Aguilar, Elena. 2013. *The Art of Coaching: Effective Strategies for School Transformation*. San Francisco, CA: Jossey-Bass.

Anderson, Lorin W., and David R. Krathwohl. 2001. *A Taxonomy for Learning, Teaching, and Assessing: A Revision of Bloom's Taxonomy of Educational Objectives*. Boston: Pearson Education Group.

Bailey, Kim, and Chris Jakicic. 2012. *Common Formative Assessment: A Toolkit for Professional Learning Communities at Work*. Bloomington, IN: Solution Tree Press.

Banks, James A., Cherry A. McGee Banks, Carlos E. Cortes, Carole L. Hahn, Merry M. Merryfield, Kogila A. Moodley, Stephen Murphy-Shigematsu, Audrey Osler, Caryn Park, and Walter C. Parker. 2005. *Democracy and Diversity: Principles and Concepts for Educating Citizens in a Global World*. Seattle, WA: Center for Multicultural Education.

Bennett, Samantha. 2007. *That Workshop Book: New Systems for Classrooms that Read, Write, and Think*. Portsmouth, NH: Heinemann.

Black, Paul, and Dylan Wiliam. 2010. "Inside the Black Box: Raising Standards Through Classroom Assessment." *Phi Delta Kappan* 92 (1): 81–90.

Brookhart, Susan M. 2008. *How to Give Effective Feedback to Your Students*. Alexandria, VA: Association for Supervision and Curriculum Development.

———. 2013. *How to Create and Use Rubrics for Formative Assessment and Grading*. Alexandria, VA: Association for Supervision and Curriculum Development.

Brooks, Jaqueline G., and Martin G. Brooks. 1999. *In Search of Understanding: The Case for Constructivist Classrooms*. Alexandria, VA: Association for Supervision and Curriculum Development.

Collier, Lorna. 2008. "The 'C's of Change': Students—and Teachers—Learn 21st Century Skills." *The Council Chronicle* 18 (2): 6–9.

Conley, David T. 2011. "Building on the Common Core." *Educational Leadership* 68 (6): 16–20.

Cornelius-White, Jeffrey H. D., and Adam P. Harbaugh. 2010. *Learner-Centered Instruction: Building Relationships for Student Success.* Thousand Oaks, CA: SAGE Publications.

Crawford, James. 2004. *Educating English Learners: Language Diversity in the Classroom,* 5th ed. Los Angeles: Bilingual Educational Services, Inc.

Cruey, Greg. 2006. "Response to Intervention: A New Model for Identifying Disabilities." https://suite.io/greg-cruey/1nd25a.

Dacey, Linda, Jayne Bamford-Lynch, and Rebeka Eston Salemi. 2013. *How to Differentiate Your Math Instruction: Lessons, Ideas, and Videos with Common Core Support, Grades K–5.* Sausalito, CA: Math Solutions.

Darling-Hammond, Linda, and John Bransford, with Pamela LePage, Karen Hammerness, and Helen Duffy, eds. 2005. *Preparing Teachers for a Changing World: What Teachers Should Learn and Be Able to Do.* San Francisco, CA: Jossey-Bass.

Davis, Robert B. 1992. "Understanding 'Understanding.'" *Journal of Mathematical Behavior* 11 (3): 225–241.

Dean, Ceri B., and Judy E. Florian. 2001. *Mathematics Standards in Classroom Practice: Standards in Classroom Practice Research Synthesis.* Aurora, CO: Mid-continent Research for Education and Learning.

Dean, Ceri B., Elizabeth Ross Hubbell, Howard Pitler, and B. J. Stone. 2012. *Classroom Instruction That Works: Research-Based Strategies for Increasing Student Achievement,* 2nd ed. Alexandria, VA: Association for Supervision and Curriculum Development.

Dueck, Myron. 2014. *Grading Smarter Not Harder: Assessment Strategies That Motivate Kids and Help Them Learn.* Alexandria, VA: Association for Supervision and Curriculum Development.

DuFour, Richard, Rebecca DuFour, Robert Eaker, and Gayle Karhanek. 2010. *Raising the Bar and Closing the Gap: Whatever It Takes.* Bloomington, IN: Solution Tree Press.

DuFour, Richard, Rebecca DuFour, Robert Eaker, and Thomas Many. 2010. *Learning by Doing: A Handbook for Professional Learning Communities at Work.* Bloomington, IN: Solution Tree Press.

Dweck, Carol S. 2008. *Mindset: The New Psychology of Success.* New York, NY: Ballantine Books.

Florida Department of Education. 2003. "FCAT 2004 Sample Test Materials." Tallahassee, FL: Florida Department of Education.

Frayer, Dorothy, Wayne C. Frederick, and Herbert J. Klausmeier. 1969. *A Schema for Testing the Level of Cognitive Mastery.* Madison, WI: Wisconsin Center for Education Research.

Frey, Nancy, Douglas Fisher, and Sandi Everlove. 2009. *Productive Group Work: How to Engage Students, Build Teamwork, and Promote Understanding.* Alexandria, VA: Association for Supervision and Curriculum Development.

Gersten, Russell, and Benjamin S. Clarke. 2007. "Characteristics of Students Brief." National Council of Teachers of Mathematics. http://www.nctm.org/news/content.aspx?id=11478.

Glatthorn, Allan A., and Jerry M. Jailall. 2009. *The Principal as Curriculum Leader: Shaping What Is Taught and Tested.* Thousand Oaks, CA: Corwin Press.

Gojak, Linda. 2011. *What's Your Math Problem? Getting to the Heart of Teaching Problem Solving.* Huntington Beach, CA: Shell Education.

Goodrich Andrade, Heidi. 2000. "Using Rubrics to Promote Thinking and Learning." *Educational Leadership* 57 (5): 13–18.

Grouws, Douglas A., and Kristin J. Cebulla. 2000. *Improving Student Achievement in Mathematics, Part 2: Recommendations for the Classroom.* Arlington, VA: Educational Research Services.

Guskey, Thomas R., and Jane M. Bailey. 2010. *Developing Standards-Based Report Cards.* Thousand Oaks, CA: Corwin Press.

Harvey, Stephanie, and Anne Goudvis. 2007. *Strategies That Work: Teaching Comprehension for Understanding and Engagement.* Portland, ME: Stenhouse Publishers.

Hattie, John. 2012. "Know Thy Impact." *Educational Leadership* 70 (1): 18–23.

Henderson, Anne T., Karen L. Mapp, Vivian R. Johnson, and Don Davies. 2007. *Beyond the Bake Sale: The Essential Guide to Family-School Partnerships.* New York, NY: The New Press.

Hiebert, James, and Douglas A. Grouws. 2007. "Effective Teaching for the Development of Skill and Conceptual Understanding of Number: What Is Most Effective?" National Council of Teachers of Mathematics. http://www.nctm.org/news/content.aspx?id=8448.

Hiebert, James, Thomas P. Carpenter, Elizabeth Fennema, Karen Fuson, Diana Wearne, Hanlie Murray, Alwyn Oliver, and Piet Human. 1997. *Making Sense: Teaching and Learning Mathematics with Understanding.* Portsmouth, NH: Heinemann.

Holmes, Nigel. 2007. "Two Mindsets." Nigel Holmes Explanation Graphics. http://nigelholmes.com/graphic/two-mindsets-stanford-magazine.

Jackson, Robyn R. 2009. *Never Work Harder Than Your Students and Other Principles of Great Teaching.* Alexandria, VA: Association for Supervision and Curriculum Development.

Kanold, Timothy D., ed. 2013. *Common Core Mathematics in a PLC at Work, Grades 6–8.* Bloomington, IN: Solution Tree Press.

Keeley, Page D., and Cheryl Rose Tobey. 2011. *Mathematics Formative Assessment: 75 Practical Strategies for Linking Assessment, Instruction, and Learning.* Thousand Oaks, CA: Corwin Press.

Knight, Jim. 2009. *Coaching: Approaches and Perspectives.* Thousand Oaks, CA: Corwin Press.

———. 2014. *Focus on Teaching: Using Video for High-Impact Instruction.* Thousand Oaks, CA: Corwin Press.

Larochelle, Marie, Nadine Bednarz, and Jim Garrison, eds. 1998. *Constructivism and Education.* Cambridge, MA: Cambridge University Press.

Learning-Focused. 2008. "Why Exemplary Schools Are Exemplary: Catching Kids Up with Acceleration." http://www.learningfocused.com.

Leinenbach, Marylin, and Anne M. Raymond. 1996. "A Two-Year Collaborative Action Research Study on the Effects of a 'Hands-On' Approach to Learning Algebra." Paper presented at the annual meeting of the North American Chapter of the International Group for the Psychology of Mathematics Education, Panama City, FL.

Leinwand, Steven. 2000. *Sensible Mathematics: A Guide for School Leaders.* Portsmouth, NH: Heinemann.

———. 2009. *Accessible Mathematics: 10 Instructional Shifts that Raise Student Achievement.* Portsmouth, NH: Heinemann.

Marzano, Robert J. 2000. *Designing a New Taxonomy of Educational Objectives.* Thousand Oaks, CA: Corwin Press.

Marzano, Robert J. 2003b. "What Works in Schools: Translating Research into Action." Alexandria, VA: ASCD.

Marzano, Robert J., Debra J. Pickering, Daisy E. Arredondo, Guy J. Blackburn, Ronald S. Brandt, Cerylle A. Moffett, Diane E. Paynter, Jane E. Pollock, and Jo Sue Whisler. 1997. *Dimensions of Learning: Trainer's Manual.* Alexandria, VA: Association for Supervision and Curriculum Development.

Marzano, Robert J. 2003. "Classroom Management That Works: Research Based Strategies for Every Teacher." Alexandria, VA: ASCD.

Marzano, Robert J., and Jana S. Marzano 2003. "Classroom Management That Works: Research Based Strategies for Every Teacher." Alexandria, VA: ASCD.

Marzano, Robert J., Jana S. Marzano, and Debra J. Pickering. 2003. *Classroom Management That Works: Research-Based Strategies for Every Teacher.* Alexandria, VA: Association for Supervision and Curriculum Development.

McLeod, Scott. 2005. "Technology Tools for Data-Driven Teachers." http://www.microsoft.com/education/ThoughtLeadersDDDM.mspx.

McTighe, Jay, and Grant Wiggins. 2013. *Essential Questions: Opening Doors to Student Understanding.* Alexandria, VA: Association for Supervision and Curriculum Development.

Moore, Lonnie. 2009. *The High-Trust Classroom: Raising Achievement from the Inside Out.* Larchmont, NY: Eye On Education.

Mora–Flores, Eugenia. 2011. *Connecting Content and Language for English Language Learners.* Huntington Beach, CA: Shell Education.

Moss, Connie M., and Susan M. Brookhart. 2012. *Learning Targets: Helping Students Aim for Understanding in Today's Lesson.* Alexandria, VA: Association for Supervision and Curriculum Development.

Moyer, Patricia S. 2000. "Communicating Mathematically: Children's Literature as a Natural Connection." *The Reading Teacher* 54 (3): 246–255.

National Council of Supervisors of Mathematics (NCSM). 2014. *It's Time: A Leadership Framework for Common Core Mathematics.* Bloomington, IN: Solution Tree Press.

National Council of Teachers of English (NCTE). 1993. *Elementary School Practices: Current Research on Language Learning.* Urbana, IL: National Council of Teachers of English.

National Council of Teachers of Mathematics (NCTM). 1989. *Curriculum and Evaluation Standards for School Mathematics.* Reston, VA: National Council of Teachers of Mathematics.

———. 2000. *Principles and Standards for School Mathematics.* Reston, VA: National Council of Teachers of Mathematics.

National Governors Association (NGA) Center for Best Practices, Council of Chief State School Officers (CCSSO). 2010. "Common Core State

Standards." Washington, DC: National Governors Association Center for Best Practices, Council of Chief State School Officers. www.corestandards.org.

National Research Council. 2001. *Adding It Up: Helping Children Learn Mathematics.* Washington, DC: National Academies Press.

National Training Laboratories. 2003. "Learning Pyramid." Bethel, MD: National Training Laboratories.

O'Neill, Jan, and Anne Conzemius. 2005. *The Power of SMART Goals: Using Goals to Improve Student Learning.* Bloomington, IN: Solution Tree Press.

Partnership for 21st Century Skills (P21). 2011. "21st Century Student Outcomes and Support Systems." http://www.P21.org.

Pearson, David P., and Margaret C. Gallagher. 1983. "The Instruction of Reading Comprehension." *Contemporary Educational Psychology* 8: 317–344.

Pearson, David P., Laura R. Roehler, Janice A. Dole, and Gerald G. Duffy. 2007. "Developing Expertise in Reading Comprehension: What Should Be Taught and How Should It Be Taught." In *Strategies That Work: Teaching Comprehension for Understanding and Engagement*, edited by Stephanie Harvey and Anne Goudvis, 17–19. Portland, ME: Stenhouse.

Pettig, Kim L. 2000. "On the Road to Differentiated Practice." *Educational Leadership* 58 (1): 14–18.

Polya, George. 1945. *How to Solve It.* Princeton, NJ: Princeton University Press.

Rendon, Sharon. 2010. "Launch, Explore, Summary Planning Tool." Handout presented at a Rapid City Area Schools training, Rapid City, SD, August.

Routman, Regie. 2008. *Teaching Essentials: Expecting the Most and Getting the Best from Every Learner, K–8.* Portsmouth, NH: Heinemann.

Rowe, Mary Budd. 1986. "Wait Time: Slowing Down May Be a Way of Speeding Up!" *Journal of Teacher Education* 37 (1): 43–50.

Sammons, Laney. 2010. *Guided Math: A Framework for Math Instruction.* Huntington Beach, CA: Shell Education.

Schrock, Connie, Kit Norris, David K. Pugalee, Richard Seitz, and Fred Hollingshead. 2013. *NCSM Great Tasks for Mathematics K–5: Engaging Activities for Effective Instruction and Assessment that Integrate the Content and Practices of the Common Core State Standards for Mathematics.* Denver, CO: National Council of Supervisors of Mathematics.

Schwahn, Charles, and Beatrice McGarvey. 2012. *Inevitable: Mass Customized Learning, Learning in the Age of Empowerment*. San Bernardino, CA: CreateSpace Independent Publishing Platform.

Seeley, Cathy L. 2004. "Engagement as a Tool for Equity." http://www.nctm.org/about/content.aspx?id=916.

———. 2014. *Smarter Than We Think: More Messages About Math, Teaching, and Learning in the 21st Century*. Sausalito, CA: Math Solutions.

Smith, Jeffrey. 1998. "Graphing Calculators in the Mathematics Classroom." Eric Digest, 1–8. Columbus, OH: ERIC Clearinghouse for Science, Mathematics, and Environmental Education.

Smith, Margaret S., Elizabeth K. Hughs, Randi A. Engle, and Mary Kay Stein. 2009. "Orchestrating Discussion." *Mathematics Teaching in the Middle School* 14 (9): 548–556.

Sousa, David A. 2006. *How the Brain Learns*, 3rd ed. Thousand Oaks, CA: Corwin Press.

Stiggins, Rick J., Judith A. Arter, Jan Chappius, and Steve Chappius. 2007. *Classroom Assessment for Student Learning: Doing It Right, Using It Well*. Upper Saddle River, NJ: Prentice Hall.

Sullo, Bob. 2009. *The Motivated Student: Unlocking the Enthusiasm for Learning*. Alexandria, VA: Association for Supervision and Curriculum Development.

Sutton, John, and Alice Krueger. 2002. *EDThoughts: What We Know About Mathematics Teaching and Learning*. Aurora, CO: Mid-continent Research for Education and Learning.

Sweeney, Diana. 2011. *Student Centered Coaching: A Guide for K–8 Coaches and Principals*. Thousand Oaks, CA: Corwin Press.

Tapper, John. 2012. *Solving for Why: Understanding, Assessing, and Teaching Students Who Struggle with Math*. Sausalito, CA: Math Solutions.

Teachers Development Group. 2011. "Mathematical Habits of Mind and Interaction." http://www.teachersdg.org.

Tomlinson, Carol Ann, and Marcia B. Imbeau. 2010. *Leading and Managing a Differentiated Classroom*. Alexandria, VA: Association for Supervision and Curriculum Development.

Tschannen-Moran, Megan. 2004. *Trust Matters: Leadership for Successful Schools*. San Francisco, CA: Jossey-Bass.

Van de Walle, John, and Lou Ann H. Lovin. 2006. *Teaching Student-Centered Mathematics: Grades 5–8.* Boston: Pearson.

Ward Hoffer, Wendy. 2012. *Minds on Mathematics: Using Math Workshop to Develop Deep Understanding In Grades 4–8.* Portsmouth, NH: Heinemann.

Webb, Norman L., et al. 2005. "Web Alignment Tool." Wisconsin Center of Education Research, University of Wisconsin-Madison. http://www.wcer.wisc.edu/WAT/index.aspx.

Wellman, Bruce, and Laura Lipton. 2004. *Data-Driven Dialogue: A Facilitator's Guide to Collaborative Inquiry.* Arlington, MA: MiraVia.

Wiggins, Grant. 2012. "Seven Keys to Effective Feedback." *Educational Leadership* 70 (1): 10–16.

Wiggins, Grant, and Jay McTighe. 2005. *Understanding by Design,* 2nd ed. Alexandria, VA: Association for Supervision and Curriculum Development.

Wiliam, Dylan. 2007. "Key Strategies Brief: Five 'Key Strategies' for Effective Formative Assessment." http://www.nctm.org/news/content.aspx?id=11474.

Woodward, John, Sybilla Beckmann, Mark Driscoll, Megan Franke, Patricia Herzig, Asha Jitendra, Kenneth R. Koedinger, Philip Ogbuehi. 2012. "Improving Mathematical Problem Solving in Grades 4 through 8: A Practice Guide (NCEE 2012–4055). Washington, DC: National Center for Education Evaluation and Regional Assistance, Institute of Education Sciences, U.S. Department of Education. http://ies.ed.gov/ncee/wwc/publications_reviews.aspx#pubsearch/. Wormeli, Rick. 2006. *Fair Isn't Always Equal: Assessing and Grading in the Differentiated Classroom.* Portland, ME: Stenhouse Publishers.

Zemelman, Steve, Harvey Daniels, and Arthur Hyde. 2005. *Best Practice: Today's Standards for Teaching and Learning in America's Schools,* 3rd ed. Portsmouth, NH: Heinemann.

Zwiers, Jeff, and Marie Crawford. 2011. *Academic Conversations: Classroom Talk That Fosters Critical Thinking and Content Understandings.* Portland, ME: Stenhouse Publishers.

Notes

Notes